Why Numbers Count

Why Numbers Count

Quantitative Literacy for Tomorrow's America

edited by
Lynn Arthur Steen

College Entrance Examination Board
New York, 1997

Founded in 1900, the College Board is a national, nonprofit membership association of schools, colleges, and other educational organizations working together to help students succeed in the transition from school to college. The Board meets the diverse needs of schools, colleges, educators, students and parents through the development of standards of excellence; by providing programs and services in guidance, assessment, admission, placement, financial aid, and teaching and learning; and by conducting forums, research, and public policy activities. In all of its work, the Board promotes universal access to high standards of learning, equity of opportunity, and sufficient financial support so that every student has the opportunity to succeed in college and work.

Figures 7–10 in John A. Dossey's "National Indicators of Quantitative Literacy" reproduced by authority of the Minister of Industry, 1997, Statistics Canada, *Literacy, Economy and Society: Results of the First International Adult Literacy Survey (1995),* Catalogue No. 89–545, pp. 41, 43, 49, and 50.

Figure 1 in Arnold Packer's "Mathematical Competencies that Employers Expect" reproduced by permission of Aon Consulting, Inc.

Library of Congress Catalog Number: 97-66326
International Standard Book Number 0-87447-503-1 (hardcover)
0-87447-577-5 (paperback)

Printed in the United States of America.

Cover design by Patricia Ryan

Contents

Authors

George W. Cobb is professor of statistics at Mount Holyoke College where he has served as department chairman and dean of studies. He is currently chair of a Joint Committee on Undergraduate Statistics of the American Statistical Association and Mathematical Association of America and director of project STATS (Statistical Thinking and Teaching Statistics). In 1993 he was elected a Fellow of the American Statistical Association.

Peter J. Denning is associate dean for computing and chair of the Computer Science Department at George Mason University. He is also director of the Center for the New Engineer, which he founded at GMU in August 1993. Previously, he was founding director of the Research Institute for Advanced Computer Science at the NASA Ames Research Center and President of the Association for Computing (ACM) from 1980-82. He currently chairs the ACM publications board, and in 1996 received ACM's Outstanding Educator Award.

John A. Dossey is distinguished university professor of mathematics at Illinois State University. He has served as president of the National Council of Teachers of Mathematics, chair of the Conference Board of the Mathematical Sciences, and chair of the Mathematical Sciences Advisory Board to the College Board. He has been involved in the National Assessment of Educational Progress in mathematics and international comparative studies of mathematics education.

Judith S. Eaton is chancellor of the Minnesota State Colleges and Universities. Previously, she served as president of the Council for Aid to Education in New York City and vice president at the American Council on Education in Washington, DC. She has held a number of administrative and faculty positions, including college presidencies at the Community College of Philadelphia and the Community College of Southern Nevada.

Susan L. Forman is professor of mathematics at Bronx Community College of the City University of New York. During 1995-97 she was senior program officer for Education at the Charles A. Dana Foundation. Previously she served as director of postsecondary programs at the Mathematical Sciences Education Board of the National Research Council and as a program officer at the Fund for the Improvement of Postsecondary Education (FIPSE).

Iddo Gal, a cognitive psychologist, currently teaches in the Department of Human Services and the School of Education at the University of Haifa, Israel. Until recently he was a research assistant professor at the University of Pennsylvania, where he directed the Adult Numeracy Project at the National Center on Adult Literacy, as well as several NSF-funded projects related to mathematics and statistics education.

Gary Hoachlander is president of MPR Associates, Inc., a consulting firm specializing in management, planning, and research. Hoachlander's own focus is on education and work, including curriculum integration, accountability, and market outcomes. He holds a Ph.D. in City and Regional Planning from the University of California, Berkeley.

Gina Kolata, a science writer for the *New York Times,* covers primarily medicine, mathematics, and biology. After studying mathematics and molecular biology in graduate school, she began writing for *Science* magazine. In 1987, after 14 years at *Science,* Kolata came to the *New York Times.*

Shirley Malcom is director of Education and Human Resources programs at the American Association for the Advancement of Science, including projects for underrepresented groups and those in the public understanding of science. A former high school and college teacher of biology, Malcom also serves on numerous boards and advisory committees, including the National Science Board and the President's Committee of Advisors on Science and Technology.

Robert Orrill is executive director, Office of Academic Affairs, the College Board, where he has been involved in issues affecting the spectrum of education as reflected in both the *Academic Preparation for College Series* and the *Thinking Series* of publications for K-12 educators, as well as scholarly titles such as *The Future of Education: Perspectives on National Standards in America* and *The Condition of American Liberal Education: Pragmatism and a Changing Tradition.* The Forum on Standards and Learning, a national subject-matter collaboration addressing questions of overall purpose and coherence in the secondary education curriculum, was founded under his aegis.

Arnold Packer is an engineer-turned-economist at the Institute for Policy Studies at Johns Hopkins University. Previously, he has served in the Office of Management and Budget, as chief economist for the Senate Budget Committee, assistant secretary of Labor, and executive director of the Secretary's Commission on Achieving Necessary Skills (SCANS). He is co-author of *Workforce 2000* (1987) and *School-to-Work* (1996).

Henry O. Pollak is visiting professor at Teachers College, Columbia University. Previously, he served as director of the Mathematics and Statistics Research Center of AT & T Bell Labs. He has served as chairman of the advisory board of the School Mathematics Study Group (SMSG), chairman of the NSF's Advisory Committee for Science Education, and as a founding member of the Board of Trustees of the North Carolina School for Science and Mathematics.

Theodore M. Porter is professor of history, specializing in history of science, at the University of California at Los Angeles. He is author of *The Rise of Statistical Thinking, 1820–1900* (1986) and *Trust in Numbers: The Pursuit of Objectivity in Science and Public Life* (1995). He has received a Guggenheim award, as well as grants from the National Science Foundation.

Glenda D. Price is provost at Spelman College in Atlanta, Georgia, with responsibilities for all faculty and curriculum issues. Previously, Price was dean of the School of Allied Health Professions at the University of Connecticut. Price is a certified laboratory scientist and allied health educator.

F. James Rutherford is chief education officer for the American Association for the Advancement of Science and director of Project 2061, a nationwide project intended to develop goals, benchmarks, and reform tools for K-12 science education. Previously, Rutherford served as director of Education at the National Science Foundation, and as professor of Science Education at New York University.

Lynn Arthur Steen is professor of mathematics at St. Olaf College in Northfield, Minnesota. Previously he served as executive director of the Mathematical Sciences Education Board, as president of the Mathematical Association of America, and as chairman of the Council of Scientific Society Presidents. He is the author of numerous books and papers on mathematics and mathematics education.

Deborah Wadsworth is executive vice president of Public Agenda, a nonpartisan, non-profit research organization, where she works to improve the quality of public deliberation. A former college administrator, she has served as program officer of the John and Mary R. Markle Foundation and as executive director of the Smart Family Foundation, where she developed a program to attract exceptional graduates of liberal arts institutions to the profession of teaching. She serves on numerous boards and committees dealing with education and public policy.

Foreword

ROBERT ORRILL
The College Board

The philosopher John Dewey said that all reflective inquiry starts from a problematic situation. These situations arise, he wrote, when a "sudden change" in experience "perplexes and challenges the mind so that it makes belief ... uncertain." Taken in its entirety, this book is about just such a situation. In the words of its editor, Lynn Steen, it is about how every one of us in these times is "caught in a rising tide of numbers," in an ever-growing flood of quantitative information that increasingly assaults us from many directions at once. It is also about the profound consequences of this event for all of us and what actions we must take if we hope, as we must, to turn this situation to advantage rather than loss.

This situation is both suddenly upon us and long in the making. It began to emerge perceptibly more than two centuries ago, but it has been enormously accelerated in recent times by the advent and powerful application of computers in gathering, processing, and disseminating information. The resulting changes are pervasive, reaching deep not just into "the environment in which we live and work, but also the entire framework of civic life." Moreover, the evolutionary pace of this change often seems bewildering. While we try to pause and reflect, Steen observes, "the relentless quantification of society continues unabated."

The news that comes with this turn of events can be either very good or very bad. At present, it's both. For those who are quantitatively literate, the new conditions are extraordinarily empowering. This can be true of anyone. The computer, Steen writes, "gives the power of numbers to ordinary citizens." Potentially, for instance, anyone now can download and analyze the entire federal budget. This kind of ready access to information heretofore almost unobtainable would seem to place more power in the hands of individuals and stimulate democratic activity. But for those that lack the "new literacy" or "numeracy" needed to take command of this situation, the consequences are starkly different. For them, the results are disabling. Tragically, Steen warns, "an innumerate citizen today is as vulnerable as the illiterate peasant of Gutenberg's time."

This warning about the possibility of a return to pre-Enlightenment conditions is profoundly troubling. Such a reversal is undesirable at any time and absolutely unacceptable in a democracy. But how can it be avoided? What actions should we take? Moreover, what exactly is "quantitative literacy" in today's world? How do we define or describe it in such a way that actions can be formulated to foster it? In seeking answers to these difficult questions, the College Board was fortunate that Lynn Steen agreed to assemble a number of relevant points of view and edit them for publication. It is now our pleasure to bring these essays and dialogues, including Lynn Steen's valuable introduction, to the American public.

It would be inappropriate in this foreword to attempt a survey of the contributions included in this volume. Lynn Steen has performed this service very well in the introduction. It does deserve special mention, however, that in identifying contributors we deliberately sought individuals who consume and use quantitative information in their professional and public lives. As a result, this volume includes voices from business and industry, the media, science, management, and public policy, among other fields. Mathematics educators respond at the close of the book, but the overall intention has been to broaden conversation and increase understanding by seeking insight from the world of work and practice. This seemed to us a necessary undertaking in a time when there is much uncertainty about how completely school mathematics matches up with the breadth of knowledge and skill most believe should make up a definition of quantitative literacy suited to complex modern conditions.

It is perhaps not surprising, therefore, that the authors and respondents in this volume come back again and again to the subject of education. How do we nurture quantitative literacy throughout all parts of American society if not through our schools and colleges? When contributors make this assumption, however, it becomes clear that their observations about quantitative literacy challenge traditional approaches usually taken to orienting the mathematics curriculum in this country. Generally speaking, there have been two such overarching orientations in this century, each of which has been pursued not just separately, but often in opposition to the other. One of these guiding perspectives conceives of mathematics as a "discipline" or "study" possessed of its own formal structure and progression. In this conception, the orientation for study at any point is not in the world outside schools or colleges, but primarily the next course in the sequence. As translated into school practice, this closed structure and its sequential order—arithmetic, algebra, geometry, trigonometry, calculus, and so on—has been largely shaped by the defining concerns and interests of research universities. This orientation, of course, brings with it the problem that the mathematics that many study in school and college is defined much too narrowly and with little regard for the needs and interests of large numbers of students.

Different from the first, the second orientation attempts to shape the mathematics curriculum according to operations and skills required in particular "vocations" or "jobs." In this view, the assumed career terminus or destiny of a student is thought of as something such as nurse, electrician, or hotel manager. It also assumes that school mathematics should be largely defined by work and economic considerations, not the liberal or theoretical learning pursued in colleges and universities. This approach, of course, has been plagued by the fact that the vocations of one generation are not those of the next, at least not in a society like ours where change is more the norm than the exception. Indeed, we are told that most individuals now change careers several times within one lifetime. Added to this, the nature of work itself currently appears to be undergoing dramatic change, marked by the need for sophisticated quantitative skills in an ever larger number of vocations. Such an approach, then, seeks stable guidance from a source or direction always in flux and undergoing change. Moreover, it would seem to be an orientation that helps students only toward fixed places in which few will remain over their entire careers. Instead of this kind of preparation, what most require are the competencies and virtuosity needed to do well in many different kinds of situations, not to mention the capacity to consider the purpose and significance of all that they do. Furthermore, as we sadly know, the dichotomizing of these dual aims (with their separation into different curricular "tracks") has helped perpetuate social and class distinctions that should have no place in American society. This, of course, is not the fault of educators but rather the result of biases deeply ingrained in our society at large.

Instead of this kind of preparation, proposals about reform of mathematics education have often centered on how to unify or combine these two orientations into one integrated approach. Especially recently, the hope has been to provide all students with a mathematical grounding that is at once theoretical and applied, conceptual and concrete, scholastic and practical. If this were achieved, it would indeed be an important step forward in meeting the educational needs described by the contributors to this book. A very real difficulty, however, may be that all this cannot be done *only* in mathematics classrooms, which does not mean, of course, that mathematics educators should put aside efforts to achieve a better balance of the two aims. Instead, it may be that the challenges involved in producing a quantitatively literate society are so many and varied that we can only hope to meet them if the responsibility is shared by teachers in all subjects. Indeed, if the needs for quantitative competence are now as pervasive in American life as this volume indicates, as well as so diverse in form and substance, it seems only common sense that the responsibility for fostering quantitative literacy should be spread broadly across the curriculum.

A clear message heard from several contributors to this book, in fact, is that opportunities to practice and utilize quantitative skills must be part of all subjects and under the assumed care of teachers in all disciplines.

But this is not only a question for schools and colleges. This book makes it evident that many voices can and should be part of the conversation, not just educators and mathematicians. Making this possible, however, requires that we organize and pursue discussion across many social and professional boundaries. This, of course, is not a comfortable conclusion to draw. As with all complex societies, we like to assign our problems to specialists and hope, if not always expect, that they will be fixed. As it turns out, though, the challenge of achieving quantitative literacy broadly in our society is one that perplexes experts and forces them to look for help outside of their own specialized communities. Despite the discomfort that comes with uncertainty, this has to be an encouraging sign. It is, as Dewey said, the necessary beginning of the kind of associative and cooperative thinking that is required for resolving any problematic situation. Our belief is that this volume makes a useful and even significant start toward bringing together the many constituencies and points of view that must be part of any such solution. Among a great many other things, what this book tells us is that all of us own the issue of quantitative literacy and all have something helpful to say about it.

* * *

All books are cooperative ventures, and *Why Numbers Count* is no exception. For helping to see this one through, several individuals deserve acknowledgment and thanks. Lynn Steen started us out and then kept us on course. Susan Forman offered advice and encouragement just when it was most needed. Dorothy Downie, Madelyn Roesch, and Joanne Daniels provided the daily help without which headway could not have been made. Geoffrey Kirshner did the same. To them and to Donald Stewart, I want to express my appreciation for the support they have given and the good work they have done in bringing this publication to completion.

Preface: The New Literacy

LYNN ARTHUR STEEN
St. Olaf College

As information becomes ever more quantitative and as society relies increasingly on computers and the data they produce, an innumerate citizen today is as vulnerable as the illiterate peasant of Gutenberg's time.

Ever since computers opened the floodgates of the information age, we have been caught in a rising tide of numbers. The ascendancy of quantitative information has changed profoundly not only the environment in which we live and work, but also the entire framework of civic life. Headlines proclaim deficit projections and unemployment numbers, holiday deaths and political polls. Editorialists debate the impact of unemployment figures on stock market trends, the cost savings of managed health care, and the impact of estrogen supplements on breast cancer rates. Behind the scenes, the mechanisms of everyday life depend increasingly on digital technology—from cellular phones to ATM machines, from bar codes to the World Wide Web.

As with hundreds of other matters small and large that command our daily attention, these constructs of modern civilization depend at their deepest level on *quantitative* information. Does the Dow Jones average really measure the state of the economy? How can we best evaluate the quality of health care? Do polls influence or just predict how people will vote? How strong is the evidence linking estrogen supplements to increased rates of breast cancer in older women? Do the new bar code readers make more or fewer mistakes than trained checkout clerks? How secure are bank accounts from ATM fraud?

Numeracy is the new literacy of our age. As the printing press gave the power of letters to the masses, so the computer gives the power of numbers to ordinary citizens. The entire federal budget is online, available for downloading and analysis by any person with access to a networked computer. So too are school board budgets, mutual fund values, and local used car prices. Every desktop computer includes spreadsheet software more powerful than programs available to professional accountants twenty years ago. No longer is the calculation of car loans or mortgage rates an esoteric specialty known only to bankers. Now all numerate citizens may determine for themselves the economic impact of their own decisions, and of the decisions their elected representatives are making on their behalf.

But just as literacy was relatively rare five centuries ago, so is numeracy today. Only one in ten U.S. adults can reliably solve problems that require two or more steps.[1] Even fewer can comprehend the complexity of issues that underlie clinical trials of new treatments for AIDS or think through the implications of a flat-tax system. Reports from international comparisons continue to show only mediocre performance from U.S. students.[2] Data from the SAT and the National Assessment of Educational Progress (NAEP) document major gaps between national goals and accomplishments.[3] The majority of today's high school graduates—not to mention dropouts—still lack fundamental "walking around" skills in quantitative literacy. When it comes to numeracy, our nation is still very much at risk.

Facing the Challenge

People often compare today's intractable problems with our success in landing on the moon. If we can put a man on the moon, why can't we cure cancer, or eliminate welfare, or reduce crime? The difference between these challenges lies not in their difficulty, but in the precision with which they are defined. Even though the moon is a moving target, it moves predictably; we know precisely where it will be, and we know when we have touched down. But cancer, welfare, and crime are not so clear; they are amorphous, evolving, and embedded in larger biological or social structures.

So too with quantitative literacy. Whatever this phrase may mean—and as the essays in this volume testify, it means very different things to different people—quantitative literacy is clearly an artifact of our culture. It serves in many arenas, including home, school, recreation, finance, work, testing, parenting, and citizenship. It requires a working synthesis of literacy and numeracy; it evolves with technology; and it both shapes and is shaped by society.

Regardless of name—numeracy, mathematics, quantitative literacy, or the derisive " 'rithmetic"—this kind of literacy is widely recognized as of fundamental importance. Yet beyond "the basics," there is little agreement about specific goals appropriate for tomorrow's world. No wonder, then, that we have made so little progress in achieving numeracy. The tension between easy agreement about importance and continuing confusion about goals produces an unhealthy paralysis in our nation's effort to become quantitatively literate. The chief purpose of this volume is to unfold for public discussion these diverse and often conflicting perspectives.

The authors in this volume speak from dissimilar professional experiences—industry and academia, government and education—and express contrasting views about the nature and importance of quantitative literacy. Although all have studied mathematics in school and use quantitative tools in their lives and work, their careers have been, for the most part, in what educators like to call the "real" world: outside the classroom. Thus they speak about quantitative literacy not as mathematics educators, but as mathematics consumers. Different (but equally diverse) perspectives of mathematicians and mathematics educators are provided in the concluding chapter.

An Informed People

Expectations for numeracy have changed very rapidly in the last two centuries. In 1800, the most quantitatively literate members of society were merchants, not scientists. In fact, as Theodore Porter points out in this volume, many people who today are just marginally numerate make more use of numbers than did scientists at the end of the eighteenth century. Numbers became really useful only after the French introduced the metric system in an attempt—still not fully successful—to overcome computational complexity and regional units of measure.

The advent of numbers as a staple of civil service enabled the rise of centralized administrative bureaucracies that were essential to new democratic societies. Nation-building depended on the flow of information "from the periphery to the center," which most usefully took the form of numbers that could be added, averaged, graphed, and analyzed. Indeed, the term "statistics" first arose at this time as the science of the state.[4] In post-revolutionary France and America, literacy and numeracy took on special political importance. An informed citizenry is "the only sure reliance for the preservation of liberty," Thomas Jefferson wrote to James Madison in 1787.[5] To participate in democracy, citizens needed the literacies of the state, which for the first time included numeracy.

To advance civic affairs, Jefferson relied not only on arithmetic and calculation, but even more on the mathematical legacy of deductive inference. The influence of axiomatic reasoning on the rhetorical style of the Declaration of Independence is revealed in its anchoring words: "We hold these truths to be self-evident... ." The source of civic consensus in a democracy, these revolutionists declared, is neither divine authority nor decisions of a monarch but deductive logic from "self-evident" propositions. Ever since, U.S. society has been constrained by a rule of law adjudicated by a Supreme Court whose decisions—and dissents—are highly mathematical arguments based on propositions, stipulations, deductions, exceptions, exclusions, and conclusions.

At the time of the Civil War, literacy meant only the ability to sign one's name. By World War II, literacy for conscripts was defined as a fourth-grade education; in mathematical terms, that meant basic arithmetic, what Lewis Carroll mocked as "ambition, distraction, uglification, and derision." Twenty-five years later, in launching the Great Society's war on poverty, the U.S. government defined literacy as an eighth-grade education—implying, in mathematics, a practical understanding of percentages as well as the ability to solve simple multistep problems.

Today most analysts agree that these "basics" are not enough, although few agree on just what more is needed. Mathematics educator Bob Moses speaks of algebra as "the new civil right."[6] Many educators and scientists urge higher-order thinking—open-ended problems, cooperative learning, communication skills. Worried parents urge emphasis on basic skills and argue for teaching methods that they remember from their school days. Indeed, the most abiding public concern seems to be that education proceed in an orderly and predictable fashion. As Deborah Wadsworth reports in this voume, the public emphasis on basics is not "just the basics," but "first the basics."

Yet despite increasing civic, educational, and economic incentives for quantitative literacy, evidence of innumeracy is not hard to find. Only two in five adults can figure correct change and tip from a restaurant bill, while only one in five can draw correct inferences about the length of a trip from a bus schedule. On national measures of quantitative literacy, fewer than one in ten adults score in the highest category, which itself is only comparable to the expectations of first-year algebra. Indeed, when asked a question about alternative tax plans that was remarkably similar to policy proposals under debate in the recent presidential election, only one out of twenty adults responded correctly. As we approach the twenty-first century, civic numeracy falls far short of the Jeffersonian ideal.

An Empowered Public

Numeracy is the currency of modern life. As it has grown in importance, it has also expanded in scope—from arithmetic in 1940, to percentages in 1960, to data analysis in 1980, to spreadsheets in 2000. And as its importance and scope have expanded, so have the economic and social consequences of innumeracy.

Much of the current emphasis on quantitative literacy is motivated by concerns about changes in a job market in which competition is no longer just regional or national, but global. The importance of quantitative literacy for employment is primarily a phenomenon of the post-war era, and most notably of the computer age. Quantitative methods of inventory and quality control were introduced to support the industrialization surge during World War II, and have in various forms dominated industry ever since. Today's managerial buzzwords, such as total quality management (TQM) and statistical process control (SPC), are direct descendants of methods introduced during the war effort. More recently, the spread of computers has brought in its wake an extraordinary reliance on quantitative methods in the work place, often centered on various manifestations of spreadsheets and graphical presentation software.

Biologist and National Science Board (NSB) member Shirley Malcom views quantitative literacy, particularly algebra, as a fundamental tool that can empower everyone—but especially the disenfranchised—regardless of background. As she makes clear in this volume, algebra is not just about solving equations, but about modeling the world. "To go from unknowns to knowns is a very powerful idea. We must learn to see algebra as a powerful friend with extraordinary explanatory power." In like fashion, the NCTM *Standards*[7] uses the metaphor of "mathematical power" to build the case for a three-year high school curriculum for all students. These *Standards* focus on providing all students—not just the college-bound—with the power of algebra, despite lack of agreement, even among professionals, on just what algebra is supposed to be.[8,9]

Although "algebra for all" is a valuable policy lever in the campaign to remedy innumeracy, it is but one aspect of quantitative literacy, much of which ranges far outside the traditional boundaries of school mathematics. Some goals for numeracy are drawn from the world of work, whereas others focus more on preparation for life. In this broader context, the focus of quantitative literacy would not be algebra, but whatever quantitative skills are needed to function in society. In today's world, spreadsheets may be more important than factoring polynomials, control charts more important than quadratic equations.

But even these broader skills are insufficient to meet the challenges of today's data-drenched society. Algebra, whether old (equations) or new (spreadsheets), is still about quantities—numbers and values, variables and parameters. As Glenda Price argues here, logical thinking, analysis of evidence, and statistical reasoning are far more important for engaged citizenship in the twenty-first century, than traditional algebraic and mathematical skills. The new literacy, from this perspective, is really about reasoning more than 'rithmetic: assessing claims, detecting fallacies, evaluating risks, weighing evidence.

What *Are* We Talking About?

It may be about time to attempt to clarify terms. For most people, *quantitative literacy* and *numeracy* are essentially synonyms, although the latter is more commonly used in British English. Both suggest a very close relationship with numbers and quantities, even though there is much more to mathematics than numbers. (In some circles, however, the term "quantitative literacy" has taken on a more narrow meaning, referring to particular parts of the school curriculum centered on exploratory data analysis and elementary statistics.[10])

Because numerical manipulations are less subtle than logical thinking—and more easily mechanized—many people believe that *quantitative reasoning* describes with far greater felicity what we are trying to achieve. Certainly the goals of engaged citizenship and high-performance work require constant thoughtfulness, not merely accurate calculations. They also require action, not just observation. For this reason, some advocate *quantitative practices* as a way of emphasizing purposefulness and accomplishment.

Mathematicians and mathematics educators (but few others) sometimes talk about *mathematical literacy,* leaving unclear whether this term is intended as a synonym for quantitative literacy. Since mathematics is generally recognized as being about more than number and quantity, the umbrella of "mathematical literacy" can provide a more welcome home for notions of reasoning (logic), space (geometry), and data analysis (statistics). But for most lay persons (as well as most mathematicians), mathematical literacy conveys a sense of advanced accomplishment more suitable to preprofessional purposes than to general literacy.

School mathematics also intrudes into this discussion because it is the primary source of quantitative literacy for most adults. The *de facto* achievement of mathematics education in the United States has been something approximating the traditional seventh- or eighth-grade curriculum: percentages, a bit of algebra, and simple geometry. That's all the mathematics that typical adults can remember ten years after they finish school.

There appears to be reasonable consensus among individuals of widely differing perspectives on the natural growth of numeracy from the basic arithmetic of grade school through the more sophisticated numerical reasoning of measurement, ratios, percentages, graphs, and exploratory data analysis that is now the centerpiece of middle school mathematics. Confusion over definitions and disputes about priorities emerge primarily beyond this level, at the core of high school mathematics. Many people identify quantitative literacy with the mathematics required of all high school graduates, whereas others view mathematical performance and quantitative literacy as two rather different enterprises.

What Do Others Think?

Around the world, all nations emphasize the use of mathematics in everyday life and its applications to other subjects. In addition, some countries stress its utility in employment, some its utility in promoting the power to reason, and a few (e.g., Italy, France) take as an explicit goal to promote precise use of language. Other goals found internationally are to stimulate the imagination; to stress the qualities of methodical, careful work; to learn to enjoy mathematics; and to value mathematics as a subject in its own right.[11]

One of the most influential and well-researched analyses of numeracy is *Mathematics Counts*,[12] the 1982 report of a British committee of inquiry on school mathematics. This report identified, among the mathematical needs of adult life, the role of mathematics as a means of communication and an "at homeness with numbers" that permits an understanding of information presented in mathematical terms. Mathematics, these authors argue, is useful precisely because it provides "a means of communication which is powerful, concise, and unambiguous."

In our own country, we can find any number of similar recommendations. An influential 1981 report from the Alfred P. Sloan Foundation[13] identified analytic skill as a "new liberal art" that provides "prodigious power." To ignore analytic skills is to "ignore the nature of the world in which the graduate will live." A few years later *A Nation at Risk*[14] argued that high schools should equip graduates to "apply mathematics in everyday situations." And in 1986 NAEP adopted a new definition of literacy that included two parts (document and quantitative) that are both actually components of numeracy.[15]

In 1988, John Allen Paulos's popular book *Innumeracy*[16] created instant public awareness of the issue and the word. For Paulos, innumeracy was signaled by the public's inability to deal with big numbers or to estimate the likelihood of coincidences. This kind of innumeracy leads to massive public confusion about issues such as political polling, medical consent, and safety regulations. For Paulos, numeracy requires, among other things, recognition of the "irreducibly probabilistic nature of life."

Perhaps because of increased public concern about the economy, recent statements about quantitative literacy have focused a bit more on issues with pocketbook implications. In 1991, the SCANS report[17] from the U.S. Department of Labor described "what work requires of schools" in terms (resources, information, systems, interpersonal skills, and technology) that are very different from the traditional curriculum that is organized into separate disciplines. Following this, in 1995 the National Research Council reported related goals for the mathematical preparation of the technical work force: "The type of problem-solving the work place demands and schools should emphasize is very different from the type many schools now teach."[18]

What Are Our Goals?

This broad and jagged range of goals for numeracy, both national and international, was summarized in a special issue of *Daedelus* devoted to "Literacy in America" under five different dimensions of numeracy: *practical,* for immediate use in the routine tasks of life; *civic*, to understand major public policy issues; *professional,* to provide skills necessary for employment; *recreational,* to appreciate games, sports, lotteries; and *cultural,* as part of the tapestry of civilization.[19] This categorization, though relatively comprehensive, does not so much define quantitative literacy as provide a framework for continued discussion. What remains is the hard question: establishing priorities.

One view widely supported by mathematics educators uses "quantitative literacy" as an umbrella for important aspects of mathematics. John Dossey identifies these aspects as data representation, numbers and operations, variables and relations, measurement, space and visualization, and chance. It is easy to see behind these terms the names of common mathematics courses such as statistics, arithmetic, algebra, geometry, and probability. They also parallel closely the fundamental strands of mathematics identified in a 1990 report of the National Research Council: quantity, change, dimension, shape, uncertainty.[20]

An important alternative view favors "reasoning" over "literacy," in part because calculation—the analog of reading—is no longer the primary need of citizens. Indeed, George Cobb describes quantitative reasoning as a "cognitive emulsion" of four very different kinds of thinking: logical, algorithmic, visual, and verbal. One can see in these categories echoes of the "four-fold way" that has become the mantra of recent reform efforts in mathematics education—that students need to learn to make transparent transitions among four different mathematical representations: algebraic, numerical, graphical, and verbal.[21]

Many people, especially those who focus on the role of mathematics in employment, believe that the goals of quantitative literacy need to be more specific and pragmatic. "Problem solving" is one of the banners under which pragmatism sails. Because a problem can be thought of as something you don't know how to do, by solving a problem you are really transforming the problem into something you do know how to do. As Henry Pollak points out, solving open-ended real-world problems requires responsiveness to two masters: solutions must be both mathematically defensible and useful in the real world. This requires not only the tools of quantitative literacy and quantitative reasoning, but also a wealth of external problem-solving strategies that depend on the scientific, managerial, and social contexts in which problems arise.

Peter Denning argues that for many managers and engineers, quantitative practices are more important than literacy. Practices encapsulate the actions, habits, procedures, and processes of people who actually do things, whereas literacy emphasizes the descriptions, theories, models, and rules of people who study things. Practices are messy, frequently ad hoc, sometimes inexplicable, and always evolving in harmony with technology. To empower citizens for active participation in the world in which they live and work, numeracy must include robust, action-oriented quantitative practices.

Providing Context

Too often we take for granted the practical benefits of numeracy. Since the age of Plato and Euclid, mathematics has been recognized as a formal science of patterns abstracted from real contexts that are irrelevant and distracting. Circles and squares are Platonic forms with properties derived from logic, not from the measurements made on imperfect real-world models.

The abiding mystery of mathematics, in Eugene Wigner's memorable phrase, is its "unreasonable effectiveness."[22] Quantitative properties of the world do generally conform to mathematical predictions. Moreover, the context of real-world problems shapes the practice of mathematics—by revealing patterns, imposing constraints, suggesting challenges, and foreshadowing theories. Nonetheless, deduction, not measurement, is the arbiter of truth in mathematics.

The role of context in mathematics poses a dilemma which is both philosophical and pedagogical. In mathematics itself, as Cobb observes, context obscures structure, yet when mathematics connects with the world, context provides meaning. Even though mathematics embedded in context often loses the very characteristics of abstraction and deduction that make it useful, when taught without relevant context it is all but unintelligible to most students. Even the best students have difficulty applying context-free mathematics to problems arising in realistic situations, or applying what they have learned in another context to a new situation.

Scientists are among those who often argue that quantitative literacy must be contextual. In science, as at work and in life, almost every number of interest is a quantity—a measurement with a unit and an implicit degree of accuracy. Even the numbers that dominate discussions of civic policy (e.g., indices of population, inflation, trade, crime) are numbers in contexts.

Unfortunately, in the words of F. James Rutherford, too many mathematics classrooms are "the natural home of measurement-free numbers."

The practice of separating symbols from context in mathematics instruction has also contributed to a widespread impression that what is taught in school is not what is most relevant for work. Much of the current impetus for improving quantitative literacy arises from the perception that advancement along a career path depends increasingly on quantitative and analytical skills. According to Robert Reich, former Secretary of Labor, the greatest imbalance between demand and supply is found in mid-level technical jobs that require modest levels of quantitative literacy. Indeed, many relatively elementary mathematical skills are undersupplied relative to demand in today's job market. Arnold Packer reports that earning levels are influenced most by mastery of mathematical skills normally taught no later than eighth grade. Moreover, many employees believe they could do their jobs better if they had greater facility with these kinds of quantitative skills.[23]

In recent years, the U.S. Departments of Education and Labor have supported development of occupational and industrial skill standards designed to clarify the prerequisites for entry-level jobs in a variety of industrial sectors.[24] Unfortunately, the mathematical content of these occupational standards is relatively low—primarily eighth- or ninth-grade level. While these minimal expectations conform to current employment practices, they pose a major challenge for education: how to justify and motivate secondary school mathematics when important government and business reports appear to say that eighth-grade mathematics is sufficient.

Schools, of course, must prepare students both for work and for postsecondary education. For many reasons, the traditional means of meeting these two goals simultaneously—vocational tracking—is no longer acceptable. According to Judith Eaton, programs that "stress the applied at the price of the conceptual" are no longer acceptable. "Curricular ghettos will not work in the new economic reality," comments NSB member Malcom. Schools can no longer prepare some students just for work and others for college. Today, they must prepare all students for both. Yet despite the virtually unanimous urging of educational and policy experts, schools still permit, even encourage, tracking.

In reality, modern high-performance work involves problems that require sophisticated reasoning and practice, but often only elementary mathematical skills. In contrast, the mathematics that students study to prepare for college requires advanced skills with abstract concepts deployed in simple (and simplistic) problem situations.[25] The difference in balance between these dimensions of quantitative literacy—higher-order problem solving on the one hand, advanced abstract mathematics on the other—accounts for much of the strain in education between vocational and academic programs.

A more promising strategy may be to engage students in challenging work-based environments that provide context for their academic courses. Conceive of work broadly, with long-term goals. Instead of carpentry or plumbing, introduce students to the entire construction industry; instead of electronics, focus on telecommunications. Engage significant mathematics at work. Test major topics against the key question: Does anyone pay to have this kind of problem solved? Eliminate topics that are of little value, while stressing neglected mathematical skills that are much in demand in the work place. As both Packer and Gary Hochlander make clear, a well-designed, broad-based, work-centered curriculum will not only provide a powerful context for mathematics, but can also demonstrate to parents, teachers, and students that vocational training is not incompatible with high mathematics standards.

Can Education Help?

"You know, they put those numbers in front of us and we just had to figure them out. It was never fun. It was just grueling...something you didn't really want to do."

"When I tried to do it, I felt like I had to throw up. It just wouldn't make any sense at all. I couldn't comprehend what's going on...I would just shut down completely. I wanted to learn, but I just couldn't get it."

These parents participating in California's "Family Math" project speak for millions who have suffered through mathematics in school.[26] No wonder mathematics is the butt of cartoons and a sure target for comedians. ("So long as there is math, there will always be prayer in schools.") Family Math parents want to help their children, but are paralyzed by fear of doing something wrong. "I feel frustrated because the way I was taught is not the same method that is being used to teach my daughter." "I see the books my son [in fourth grade] brings home, but most of the time I don't understand the math that is done there, so I can't help him."

Too often, numeracy in school turns out to be just a sterile version of formula appreciation. The incantation $A = \pi r^2$ carries about the same meaning for most people as does $E = mc^2$. Most of the reform efforts in mathematics education for the last half century or more have struggled with the problem of restoring meaning to mathematics. In Malcom's words, "If you change the way mathematics is taught, you'll be surprised at who can learn mathematics. The idea of fitting the subject to the audience is real uncharted territory." What we do know is that memorizing formulas doesn't make anyone literate.

Despite lip service paid to the importance of applications, school mathematics continues to emphasize school-focused goals rather than the goals expected by the outside world. Becoming numerate, according to Iddo Gal, requires much more than knowing mathematics. It involves, among other things, a synthesis of literacy and numeracy that is rarely encouraged in school and a range of contexts from computation through decision making to interpretation. Too often, school mathematics stops with computation.

So too does mathematics in college. Relatively few colleges or universities actually focus on quantitative literacy for all students. Of those that do, it is rare to find any whose expectations exceed what the NCTM now recommends for all high school graduates. Both the Mathematical Association of America (MAA) and the American Mathematical Association of Two-Year Colleges (AMATYC) have recently issued reports[27, 28] containing recommendations for quantitative literacy in undergraduate courses, but very few institutions have programs that meet these guidelines. Students need multiple opportunities and multiple inducements to develop their quantitative skills. As Price makes clear in this volume, these skills cannot be mastered in any one course, but, like writing, must be reinforced by repeated use in many courses and not just offered by mathematics departments.

Why Numbers Count

The relentless quantification of society continues unabated. The tendency to reduce complex information to a few numbers is overwhelming—in health care, in social policy, in political analysis, in education. The production of uniformity through numbers seems to make the world more administrable, although not always more fair. As numbers shape public policy, they take on, in Porter's words, "totemic significance." Although the widespread availability of data should enrich public discourse, inevitable oversimplifications and misinterpretations may ultimately cheapen it. The power and ubiquity of computers make it possible for many people to calculate and communicate numbers they do not understand. Instead of enhancing Jeffersonian democracy, limited numeracy can easily shift the balance to a technocracy.

Innumeracy hurts in other ways as well. For example, public policy issues may increasingly move beyond the intellectual grasp of citizens who lack appropriate skills in quantitative reasoning. As Gina Kolata argues here, innumeracy encourages the view that all opinions are equally valid, that whenever there is disagreement the truth lies somewhere in the middle. Innumeracy thus becomes another means of disenfranchisement: by reinforcing the idea that truth is relative and unknowable, people with the least defenses against charlatans will be most vulnerable.

Innumeracy also perpetuates welfare, harms health, and weakens families. Without requisite quantitative skills, individuals will find it very difficult to make a transition from welfare to work. Without critical skills to assess medical options, individuals will often fall victim to false claims and questionable treatments. Without the skills to manage a household budget, many become victims of easy credit or consumer fraud. In short, an innumerate citizen today is as vulnerable as the illiterate peasant of Gutenberg's time.

As the technology of literacy advances from ink on paper to images in cyberspace, the practices of literacy take on multiple facets reflecting the multiple modes of modern communication. Literacy is no longer just a matter of words, sentences, and paragraphs, but also of data, measurements, graphs, and inferences. Pattern and number lurk behind words and sentences, in machines and computers, in organizations and networks. Literacy is about reading and reasoning, writing and calculating; it is about solving problems and using technology; it is about practices as well as knowledge, procedures as well as concepts. Numbers count because ideas count.

References

1. National Center for Educational Statistics. *Adult Literacy in America.* Washington, DC: U.S. Department of Education, 1993.
2. International Education Association. *Mathematics Achievement in the Middle School Years: IEA's Third International Mathematics and Science Study.* Boston, MA: TIMSS International Study Center, 1996.
3. National Assessment of Educational Progress. *1994 Trends in Academic Progress.* Washington, DC: National Center for Education Statistics, 1996.
4. Victor Cohn. *News and Numbers.* Ames, IA: Iowa State University Press, 1989.
5. Adrienne Koch and William Peden. *The Life and Selected Writing of Thomas Jefferson.* New York: Random House, 1993, p. 407.
6. Alexis Jetter. "Mississippi Learning." *New York Times Magazine.* Feb. 21, 1993, 21.
7. National Council of Teachers of Mathematics. *Curriculum and Evaluation Standards for School Mathematics.* Reston, VA: National Council of Teachers of Mathematics, 1989.
8. Carole Lacampagne, William Blair, and Jim Kaput. *The Algebra Initiative Colloquium* (2 Vols.). Washington, DC: U. S. Department of Education, 1995.
9. Zalman Usiskin. "Why is Algebra Important to Learn?" *American Educator* (Spring 1995) 30-37.
10. Jerry Moreno (Editor). *The Statistics Teacher Network.* Newsletter of the ASA/NCTM Joint Committee on the Curriculum in Statistics and Probability.
11. Geoffrey Howson. *National Curricula in Mathematics.* Leicester, England: The Mathematical Association, 1991.
12. Wilfred H. Cockroft (Editor). *Mathematics Counts.* London: Her Majesty's Stationery Office, 1982.
13. Stephen White, et al. *The New Liberal Arts: An Exchange of Views.* New York: The Alfred P. Sloan Foundation, 1981.
14. National Commission on Excellence in Education. *A Nation at Risk: The Imperative for Educational Reform.* Washington, DC: U. S. Department of Education, 1983.
15. Irwin S. Kirsch and Ann Jungeblut. *Literacy: Profiles of America's Young Adults.* National Assessment of Educational Progress. Princeton, NJ: Educational Testing Service, 1986.

16. John Allen Paulos. *Innumeracy: Mathematical Illiteracy and Its Consequences.* New York: Hill and Wang, 1988.
17. Secretary's Commission on Achieving Necessary Skills. *What Work Requires of Schools: A SCANS Reports for America 2000.* Washington, DC: U. S. Department of Labor, 1991.
18. Susan L. Forman (Editor). *Mathematical Preparation of the Technical Work Force.* Washington, DC: National Research Council, 1995.
19. Lynn Arthur Steen. "Numeracy." *Daedalus* (Spring 1990) 211-231.
20. Lynn Arthur Steen (Editor). *On the Shoulders of Giants: New Approaches to Numeracy.* Washington, DC: National Research Council, 1990.
21. Alan C. Tucker and James R. C. Leitzel. *Assessing Calculus Reform Efforts.* Washington, DC: Mathematical Association of America, 1995.
22. Eugene P. Wigner. "The Unreasonable Effectiveness of Mathematics in the Natural Sciences." *Communications on Pure and Applied Mathematics.* 13 (February 1960) 10-14.
23. Irwin S. Kirsch, Ann Jungeblut, and Anne Campbell. *Beyond the School Doors: The Literacy Needs of Job Seekers Served by the U.S. Department of Labor.* Princeton, NJ: Educational Testing Service, 1992.
24. Thomas Bailey and Donna Merritt. *Making Sense of Industry-Based Skill Standards.* Berkeley, CA: National Center for Research in Vocational Education, 1995.
25. Susan L. Forman and Lynn Arthur Steen. "Mathematics for Work and Life." In *Prospects for School Mathematics.* Iris M. Carl, Editor. Reston, VA: National Council of Teachers of Mathematics, 1995, pp. 219-241.
26. Virginia Thompson. *Family Math.* Project EQUALS. Berkeley, CA: Lawrence Hall of Science, 1995 (Unpublished interview transcripts).
27. Linda Sons (Editor). *Quantitative Reasoning for College Literacy.* Washington, DC: Mathematical Association of America, 1996.
28. Don Cohen (Editor). *Crossroads in Mathematics: Standards for Introductory College Mathematics Before Calculus.* Memphis, TN: American Mathematical Association of Two-Year Colleges, 1995.

The Triumph of Numbers: Civic Implications of Quantitative Literacy

THEODORE M. PORTER
University of California, Los Angeles

The rise of numbers as barometers of society—through censuses, indices, taxes, elections, polls—built bureaucracies and enabled centralized government. As numbers flow from periphery to center, they gain totemic significance and unparalleled influence over ordinary lives.

As recently as two centuries ago, few people were quantitatively literate. Scientists were a partial exception, but not a very important one, because there were so few of them, and because many even in what we now consider the physical sciences made little use of numbers. Measurement and calculation had acquired some role in limited areas of engineering, such as the design of roads, waterwheels, and engines, or the analysis of ores in mining. Certainly, though, engineering was not defined by the use of mathematics, as it is today.

As recently as 1800, the most important practitioners of quantification were merchants. The manipulation of quantities was an extraordinarily challenging task in those days when measures often varied from town to town, when there were different measures for different substances, and when almost nothing was decimalized. As John Heilbron remarked in an essay on the Enlightenment: "Calculation of the price of a piece of cloth 2 yards 1 foot 4 inches square at 3 pence 2 farthings the square foot was a sufficient challenge. To change it into aunes, pieds, livres, and deniers, and to proceed to a problem in bushels and cubic king's feet, would have puzzled Archimedes."[1] Little wonder that arithmetic was the business of specialists. The complexity of measures provided one of the main sources of support for mathematicians in Europe through the eighteenth century.

The metric system, created by French scientists during the revolutionary 1790s, was an effort to overcome the computational complexity and regional specificity of measures. It was sometimes attacked in its own day, and continues to be criticized in ours, as a disruption of local custom by centralized political and economic powers and by the scientific elite. That the rationalization of measures contributed to the concentration of political power is very clear. Whether it worked to the disadvantage of peasants and craftsmen, however, may be doubted. The savants and *philosophes* who promoted it did not want to install a system that would forever seem alien to the common people, but rather thought that a rationalization of measures would improve their lot. The looseness and negotiability of the old measures worked to the advantage of territorial elites, especially noble seigneurs, who were almost universally suspected by peasants of augmenting the measures in which their dues were paid. A lack of computational ability also implied a kind of dependence. The Enlightenment stood for the greater autonomy of individuals in this as in other respects: every man his own calculator. The new names of measures, with their mostly Greek roots, made them difficult to learn, but it certainly was easier to manipulate uniform decimal quantities than a patchwork of heterogeneous ones.[2]

It would be interesting to know how often ordinary laboring people in the eighteenth century or earlier faced measuring problems that went beyond their computational powers. There is no doubt that the late eighteenth century was the beginning of a huge expansion in counting, measuring, and calculating. Only then did a spirit of quantification begin to become pervasive in the natural sciences. By this I mean that quantification was no longer limited to mathematized fields such as astronomy and mechanics, but measurement became routine in studies like chemistry and mineralogy. At the same time, engineers and natural historians measured and mapped the surface of the earth at a new level of detail and richness. The naturalist Alexander von Humboldt might be taken as exemplary: he traveled widely to mountains, deserts, and forests, always in the company of his instruments, measuring elevations, barometric pressures, and temperatures, and inscribing the numbers on maps. Exemplary also are the French state engineers who surveyed the territory in order to provide for defense against foreign invasion and to plan roads and canals.

Building Bureaucracies

The growth of quantification was evident in the social domain as well. Most national censuses in western Europe and North America were introduced between about 1790 and 1830. In the course of the nineteenth century, they became far more comprehensive and detailed. At least as important, European states formed the habit of publishing great compilations of numbers pertaining to population, births and deaths, crime, and trade. To the extent that such figures had been gathered at all in the eighteenth century, they had generally been regarded as state secrets. Here again, quantification had an important civic dimension, and many nineteenth-century social statisticians regarded their occupation as linked to democratic openness and to public participation in the political realm.

Inevitably, this did not merely expose some of the secret workings of government but tended also to change government itself. Numbers and the discourse surrounding numbers took on a life of their own. Population, health, safety, and prosperity—subtle entities whose fluctuations had once been almost invisible—were in a way made real by the process of measuring them. Soon they began to reshape public life. Bureaucracies came to be characterized more and more by the flow of information and less and less by the independent, often secretive actions of autonomous officials. The success of statesmen and of policies was judged increasingly on changes in economic, demographic, and judicial numbers.

This expansion in the collection, publication, and use of numbers, when considered as a phenomenon of the last two or three centuries, stands out as one of the really important developments of modern history. Contemporary science would be almost unthinkable without it, as would our current systems of finance, bureaucracy, regulation, engineering, medicine, insurance, trade, and tax collection, to give only a partial list. Accounts, statistics, and measurements have contributed enormously to administrative centralization and to the growth of the state and of large business enterprises.

The flow of information from periphery to center often takes the form of numbers. Only a very brave historian would claim that it had to happen this way, that large organizations and centralized governments are by their nature dependent on quantitative information. But the link is by no means merely accidental. Numbers are ideally suited to combination and abstraction, and hence to distillation into ever simpler forms. They can easily be passed along from the provinces to the capital, or up an administrative hierarchy to the political leadership, and combined with other numbers as they proceed. Often, this requires no more than adding them together. Combining numbers is much more convenient than abridging thousands of written reports, each about a different district or problem, into one brief summary.

It is, of course, possible to reject such amalgamation, even with respect to numbers. To combine the murder rate in New York City with that in the rest of the state, or to average unemployment rates of black urban teenagers with those of college-educated white adults, may be regarded as adding apples and oranges. Numerical descriptions can be gathered and reported at almost any level of detail. But the conditions that have led to our current high level of quantitative activity have also favored a strong move toward abstraction. The different patterns of crime and employment in Santa Ana and Newport Beach may be important to administrators in southern California's Orange County, but we can scarcely expect federal officials in Washington to think about such patterns when they are trying to assess the general level of security and prosperity in the United States. Neither will these local differences command national attention when the F.B.I. or the Bureau of Labor Statistics releases an annual report. Neighborhood residents are likely to decide whether it is safe to walk the streets at night, or how the economy of their community is faring, mainly on the basis of anecdotal evidence picked up in conversation and from television or radio. The numbers are not collected specifically for local populations, but are used mainly by people farther away, people who know much less about local conditions. For them, the disappearance of detail is no disadvantage, but rather is the only way they can hope to deal with what would otherwise be a flood of information.

By now numbers surround us. No important aspect of life is beyond their reach. Factories and shops, freeways and harbors, schools and concert halls, churches and bedrooms are all surveyed or sampled, then recorded and reported in quantitative terms. Often the collection of the numbers depends on great feats of bureaucratic organization. It requires not only a vast number of employees, supervisors, and bosses to collect numbers all over a country, but also detailed definitions of categories, protocols for locating people, and routines for dealing with problematic cases. If the customary organization of automobile factories, for example, differs radically between Alabama or Tennessee and Michigan, it may be very difficult to assemble detailed occupational statistics for the entire United States. The statistical bureaucracy, generally in alliance with a regulatory one, may become a powerful force for the imposition of standard categories—not just on the paper forms, but also in the real world. This is a kind of simplification through which people and products are made to fit these standard categories, the better to count them. This production of uniformity through numbers makes the world more readily administrable.

To the extent that this process yields up accurate and useful descriptive numbers, it can also promote a wider diffusion of information, which seems favorable to democracy. At least people can more easily feel informed even when they lack personal experience of the phenomena being discussed. This process tends also to enhance centralized administration. What it diminishes is the authority of elites close to the scene of action. Traditionally, local knowledge was the property of rich bourgeois, landowners, and aristocrats. The triumph of quantification has tended to disrupt their monopoly of information, and hence to reduce their power. Sometimes local power was centered around town meetings or village councils; in this case too, the growth of quantification has tended to diminish local autonomy in favor of the administrative state.

The implications of the growth and influence of quantification extend also to the quality and character of public discourse. Much of the bureaucratic apparatus of collecting numbers and producing quantitative information is designed with the purpose of informing the public without telling it too much. Efficient decision making is more and more seen to require eliminating complexities by reducing large public policy issues to just a few numbers. While those who know and understand the numbers have a certain advantage over those who do not, even the bureaucrats have only limited control over the quantitative information that is provided to the public. Nonetheless, whether or not summary numbers accurately measure what they were intended to measure or provide sufficient information for reasoned discussion and decision making on the part of the public, such numbers often take on a life of their own. Elections can be won or lost, managers promoted or fired, and share prices or currencies rise or fall, based on illusory or misleading numbers that influence the thought and behavior of the public.

Cost-Benefit Analysis

The history of cost-benefit analysis illuminates very nicely the pressures to eliminate complexities by reducing large questions to just a few numbers. Cost-benefit analysis is by now applied to public policy decisions of almost all kinds, including transportation, health, safety, education, and recreation. It was originally devised not mainly by economists, but rather by engineers working in public bureaucracies. A leading role was played by the U.S. Army Corps of Engineers, which has administered the nation's waterways for purposes of navigation almost since the founding of the Republic, and more recently became highly active in flood control. It is no accident that this form of quantitative analysis emerged from public bureaucracies rather than private industry. It is less an expression of market rationality than of political or administrative rationality.

The measurement of costs and benefits initially emerged as a form of accounting, a way of keeping track of expenditures and potential revenues. The Corps of Engineers, however, did not charge shippers for navigational improvements, and it charged little for flood control. Hence its measure of benefits could not refer to actual, or even to anticipated, future revenues, but only to an imagined, potential revenue. This was generally conceived in terms of benefits realized by those who shipped on canals or who could with greater security build or farm on a flood plain. In the early years of cost-benefit analysis, the first three decades of the twentieth century, the Corps of Engineers rarely performed these analyses with any pretension of comprehensiveness or rigor. The methods were rather loose, and the economic implications of projects were determined informally among the highest officers in the Corps bureaucracy.

The push for quantitative rigor was initiated in part by Congress, which was evidently embarrassed by the manifestly uneconomic character of some of the projects it approved. A flood-control act of 1936 specified that no flood-control project should be approved unless it showed benefits in excess of costs. This gave Congress some basis for arguing that it spent this money rationally and did not merely pander to powerful interests.

Initially, the cost-benefit criterion left much room for interpretation, and also of course for disguised political influence. But this freedom to interpret was difficult to preserve when the Corps was challenged. And it was challenged in the 1930s and especially after 1945 by electric utilities, railroads, and other agencies in the federal bureaucracy. Important decisions were made to hinge on whether the benefits of a particular dam were predominantly linked to irrigation, flood control, or power production. The sharply different methods for quantifying benefits that separated the Corps from the Bureau of Reclamation, and both from the Soil Conservation Service of the Department of Agriculture, led to no end of bickering and embarrassment. As a consequence, the rules of cost-benefit analysis had to be made uniform, so far as possible, and then specified in numbing detail. Of course it was never possible to exclude interpretation entirely, and in high-profile cases the politics could still be overwhelming. Nonetheless, a genuine effort was made to rein in and establish rules governing the exercise of political judgment.

The pressure to compare benefits with costs created strong incentives to express all elements of the analysis in monetary terms. There is much irony in this part of the story. Among the elements pertaining to flood control that now had to be quantified were scenic value and the preservation or loss of human life. The U.S. public has never been willing to view with equanimity the assignment of a fixed money value to human life. Use of values that are differentiated by age, sex, or race makes the calculation still less tolerable. Yet it was the pressure of political conflict that discredited the informal methods of a previous generation of engineers and encouraged them to express as much as possible in monetary terms.

The process of quantification was of course enormously complex, and as a result, issues such as scenic beauty, flood protection, lives saved, new construction on the flood plain, changes in water quality, recreation opportunities, power generation, and mosquito control all disappeared from view. What remained were two numbers: one standing for costs, the other for benefits. Decisions were to be based on a comparison of just these two numbers. In actual practice it is almost never quite so simple, but this was the ideal that emerged from the political conflict over water control from the 1930s to the 1950s, and that has spread to other forms of public investment and regulation since the 1960s.[3]

The Triumph of Quantification

The reduction of complex information to a few numbers, or even a single number, has become a common feature of contemporary life. It is not limited to numbers designed for presentation to the general public. Many scientists, for example, calculate levels of statistical significance as an indication of whether an observed result is real, and not merely an effect of chance. But even in science, such methods have been associated with issues of a broadly political and administrative kind, such as drug regulation and educational psychology. In public statistics, the use of a single index number to summarize a broad and sometimes elusive concept has become extremely common. Inflation is measured by indices of consumer and wholesale prices. Trade balances, worker productivity, gross domestic product, the unemployment rate, and mean household income are other indices used to assess the health of the U.S. economy or the success of current economic policy. In a similar way, crime rates and birth rates out of wedlock are used as indices of security and morality.

Specialists understand that these index numbers are subject to a host of uncertainties. Random sampling errors, the one element of the operation that can be estimated easily using mathematical probability, are often among the least important of these. There are serious systematic errors because the reporting of crime, trade, wealth, and the like is neither random nor comprehensive. Also the reports are not entirely honest. Still more important, in many cases, is an inescapable fuzziness in the concepts themselves, or at least in their capacity to represent an entity that really matters. Measures of the cost of living should perhaps, for some purposes, ignore housing prices and rely instead on rental costs. It is now widely thought that existing indices exaggerate inflation by failing to take into account improvements in the quality of products. All these numbers are subject in addition to significant regional differences.

One can go on ad nauseam in this vein. The point is not to criticize the measures, nor even to explain their limitations, but to make a more general point. These indices are the result of so much simplification that a single number may be more misleading than helpful. Reporters are becoming increasingly conscious of these problems, and in relatively sophisticated media index numbers are sometimes broken down a bit and their complex meanings explained. But in other contexts they are often assigned a totemic significance. The reporting of simplified numbers may often distort as much as it reveals. Sometimes, especially in political debate, the distortion is deliberate. The result may cheapen public discourse rather than enrich it.

Given this importance of numbers, quantitative literacy is not merely a matter of mathematical competence but has civic and political implications as well. The most obvious is that in a world full of numbers, the innumerate will be at a disadvantage. This will of course help to determine whether they can secure good jobs and whether they can work effectively within such jobs. Numeracy affects the ability of people to make informed political choices. It is not simply that innumerates will not understand some of the information reaching them from conversations, newspapers and television, but also that press treatment of quantitative issues cannot be very deep or probing if the audience does not understand how the numbers came about and what they really mean. The gap between the massive reliance on quantitative methods in business and government and the very limited numeracy of the wider public shifts the balance of authority away from democracy toward a kind of technocracy. It means also that many political choices made democratically will be less well-informed. It seems unlikely that quantitative analysis will in the near future become less prominent in political and economic life. The trend for the last roughly 200 years has been toward greater and greater quantification. Informed and empowered citizens need to know more about how quantitative analysis works, and about what index numbers mean.

Beneath and Beyond Numbers

On the whole, understanding the role of numbers in the world has not been a high priority of mathematics education. Up to high school, mathematics generally means the manipulation of quantities and symbols. Students are required to perform arithmetic calculations and to solve equations. Mathematical thinking is rarely applied in elementary and secondary education to any substantive field of inquiry other than physics. The problem of attaching numbers to the world is given rather little attention, except in relation to certain idealized problems chosen to promote a certain way of thinking rather than to learn about the world. Students are taught to calculate how many dimes and quarters they have if their pockets contain seven coins worth a dollar, or where two trains will collide if one is traveling west at 70 miles per hour and the other going east at 50. Rarely do they learn what a stratified sample is, or how an unemployment rate is determined, or what the smog index measures. The sorts of numbers that modern citizens are likely to confront in their lives as citizens and voters have little place in the modern curriculum, except in advanced college-level courses on rather specialized topics such as economic statistics, public accounting, or social surveys. Such courses are generally seen as preprofessional, not as topics for general education.

The widespread lack of quantitative literacy is much bemoaned in the United States. Cited as evidence are the low scores of students on standardized mathematics tests in comparison with students of other countries. This is itself a bit of quantitative wisdom, one that is instructive but also misleading. Supposing the result is correct so far as it goes, there remains the problem that it takes for granted a certain understanding of quantitative literacy. I would argue that achievement in the study of formal mathematics is only one element, albeit an important one, in gauging quantitative literacy. An understanding of measurement and calculation as social activities, and of the problems to which they are applied in the domains of economics, society, medicine, and bureaucratic regulation, should also be regarded as important for the education of citizens, and perhaps even as partial measures of quantitative literacy.

Mathematics is central to our modern scientific understanding of the natural and social worlds. But our reliance on it is not simply a consequence of its perceived objective validity. Quantification is also a critical element in how we conduct our affairs, exchange goods and services, define and enforce regulations, and communicate knowledge. In all these senses, the world has become much more thoroughly quantitative since 1800, and is likely to become still more quantitative in the next century. At the same time, measurement systems have become simpler and computational power has increased tremendously, so that many who think of themselves as only marginally competent in mathematics make more use of numbers than many physicists did at the time of the American and French Revolutions. Often, in fact, such people are only marginally competent, or less, so that they calculate and communicate numbers that they do not really understand.

Except for a few rather specialized purposes, an eighteenth-century man or woman did not need to be numerate—beyond a basic ability to handle money. Citizens at the end of the twentieth century have more experience with the domain of quantity, but not nearly enough to deal competently with the flood of numbers that circulate throughout the world.

The vast expansion and, to a degree, democratization of numbers has meant also that for many purposes there is a huge premium on simplification. The reduction of complex economic and social issues to index numbers and difficult political choices to a comparison of costs and benefits can be highly misleading. Quantitative literacy ought to mean an ability to see below the surface and to demand enough information to get at the real issues.

Endnotes

1. John Heilbron, "Measure of Enlightenment," in *The Quantifying Spirit in the Eighteenth Century,* eds. T. Frängsmyr, J. Heilbron, and R. Rider (Berkeley, Calif.: University of California Press, 1990), 212.
2. On the implementation and effects of the metric system, see Witold Kula, *Measures and Men,* trans. Richard Szreter (Princeton, N.J.: Princeton University Press, 1986); Ken Alder, "A Revolution to Measure: The Political Economy of the Metric System in France," in *The Values of Precision,* ed. M. Norton Wise (Princeton, N.J.: Princeton University Press, 1995), 39–71.
3. Theodore M. Porter, *Trust in Numbers: The Pursuit of Objectivity in Science and Public Life* (Princeton, N.J.: Princeton University Press, 1995).

Civic Numeracy: Does the Public Care?

DEBORAH WADSWORTH
Executive Vice President, Public Agenda

Concerns about numeracy overlook growing public resistance to technical or "expert" solutions. The issue is not just to enhance expertise, but to put first things first.

Many major national issues—for example, balanced budget, national debt, health insurance, Social Security—rest on complex quantitative arguments such as rates of growth and economic forecasts. In what ways does our collective innumeracy impede civic discourse?

Your question touches on some of the most profound and troubling issues facing this country today. At Public Agenda, all of our work is predicated on several principles. First, a number of complex issues facing our country cannot be resolved by experts and leaders alone without the support of the public. Further, we believe that just as national leaders have spent years thinking through these issues, the public, too, must go through a multiyear and multistage process of wrestling with the issues and working toward consensus or compromise. Finally, we also believe that the process of public debate and dialogue is frequently sidetracked by obstacles and resistance, which result in the frustrating climate of stalemate and wheel-spinning that all too often characterizes civic discourse.

The issues you mention—such as the budget deficit—do indeed have a technical and mathematical component that may be poorly understood by the public. Debates between experts and the public often turn on precisely these issues, and this does cause severe policy problems for our country. An obvious example of this is the current discussion of Medicare and Social Security. Experts are fond of pointing out that these programs are not really actuarially sound insurance programs and that, in most cases, the benefits an individual receives far exceed the contributions paid in over the years.

Many experts believe that this provides a justification for cutbacks in some of the benefits, as one step in balancing the federal budget. The public, however, regards these entitlements as something it has bought and paid for with insurance premiums, and regards any attempt to tamper with them as a profound violation of its rights. As a result, these entitlement programs have become what politicians call a "third rail"—"touch it and you die."

It is tempting, as your question suggests, to think of this as a problem of innumeracy, since the simplest mathematical calculation reveals the imbalance between premiums and benefits. There is, of course, a certain amount of truth in this way of characterizing the problem. But there is also a great danger in seeing the debate this way. Defining this issue in terms of public innumeracy feeds into a view that is all too common among the national leadership, and one that is deeply resented by the public. It suggests the public is stupid and the experts are smart, and if people were just a little less dumb or a little more mathematically sophisticated, we could balance the budget and solve more of our national problems.

Our work at Public Agenda, however, has attempted to frame the public's resistance to the solutions posed by experts in a broader context. We see the root problem not as a conflict between an innumerate and uneducated public versus a numerate and sophisticated elite, but as a conflict between the public's moral and value-driven perspective on issues and an expert perspective that is increasingly technical and value-free. It is not that the experts are smart and the public is dumb, but that the two groups approach issues from different and somewhat incompatible viewpoints.

To choose only one example, we have done a number of studies on the attitudes of the public and of national leaders on health care. Many leaders see the health care cost explosion as driven by technical factors such as the growing number of older people or the inflationary impact of new technologies. They look for a technical solution, therefore, in new systems (such as HMOs) to control costs. Our studies show that the public, for its part, sees the issue in moral terms. People tend to diagnose the problem as one of waste, fraud, greed, and abusive malpractice suits. Experts tend to trivialize these issues as a minor part of the problem. People who are having trouble affording health care coverage, however, are morally outraged by seeing a hospital bill that includes $5 for ice someone could get in the cafeteria for free. It is not that people are unwilling to think about other aspects of the problem; rather, they are unwilling to focus on those aspects until their moral concerns have been heard and addressed.

The problem is that experts either completely ignore the public's perspective, or, when they finally hear it, use it as a way to manipulate the debate. The media are thus much more likely to focus on outrageous examples of waste in health care than on the cost implications of new technologies. Politicians too, advised by media consultants, begin to use the public's moral outrage as a way to achieve their own goals—evoking fear and anger at "big government" or raising alarms about government violating the rights of seniors. The result is that the media are used to pump up the public's anger, cynicism, and frustration. While this may help win specific elections, it ultimately weakens the fabric of civic debate for everyone.

My point is that seeing the problem as numeracy versus innumeracy distracts us from the real issue, which is a growing gap between expert reliance on technical analysis and the public's moral perspective. The public may indeed be innumerate, but at least in part this innumeracy may grow not out of a lack of knowledge of mathematics but out of a feeling that the real problems are not technical but moral. Until experts begin to address the public's moral concerns, and until politicians and the media find better ways to engage the public on complex issues (and we have some ideas about how they might do that), the public is unlikely to be interested in thinking about complex policy issues from a technical and mathematical perspective. We will not, in other words, deal with problems such as Social Security by educating the public on the mathematics of social insurance systems. We need to find ways to address and debate the public's moral concerns and its anxieties about the future.

A Question of Priorities

Conventional wisdom holds that the United States is facing a crisis over the widening gap between high-wage, high-skill jobs and low-wage, low-skill jobs. To what degree is this gap caused by differences in quantitative and technical literacy, as many educators and politicians would have us believe? Would increased quantitative literacy really help close the gap?

Rather than answering your question directly, let me use it as an example of the expert-public gap that I mentioned before. The thinking of those national leaders who stress quantitative literacy as the answer to the problem of the disappearing middle class certainly has an appeal. We are already in a high-tech, knowledge-intensive world economy. Who could deny that the possession of technical skills is what creates the divide, and that greater quantitative literacy for more citizens could help overcome it? This view may be correct, but it is somewhat at odds with the way the public addresses the problem, and it is once again symptomatic of the growing gap between leaders and the public which we have already discussed.

As are their leaders, the public is deeply troubled both by the difficulties of young people in finding decent jobs and by the inadequacies of U.S. education. People do see a lack of quantitative knowledge and skills as part of the problem. But the public is even more preoccupied by deeper concerns at the values level, such as the need for schools that are safe and disciplined where young people can learn work habits such as being on time and, where they will learn how to deal with people in authority. What concerns the public is the perceived inability of many young people to read, write, and even speak basic English, or to conduct themselves in an appropriate manner in the adult world. In *First Things First,* a study of public attitudes toward the schools, we found that two approaches to improving the schools received nearly universal approval and topped all other factors. They were: "Not allowing kids to graduate from high school unless they clearly demonstrate they can write and speak English well," and "Emphasizing such work habits as being on time, dependability, and discipline." Both were supported by nearly 9 out of 10 Americans (88 percent). By contrast, only 37 percent said it is absolutely essential for schools to teach advanced mathematics such as calculus.

It is not that the public denies the importance of quantitative literacy. The debate is about priorities. People are morally outraged by their perceptions that schools are violent and unsafe places for children to learn; that a few troublemakers are allowed to disrupt classrooms so that no one can learn; and that young people can graduate from school without any idea about how to behave in the adult world and without even minimal English competency. In the public's eye, these things are holding both our kids and our economy back, and it would like to see attention paid to these priorities even before issues concerning academic competence. When national leaders talk about quantitative literacy without addressing the public's main concerns at the values level, people become deeply frustrated, not because what the leaders are saying is false, but because their talk reflects a lack of interest in even more urgent priorities. Discussions of quantitative literacy, if they are advanced in isolation from the context of broader moral and value concerns, are more likely to alienate the public than to win its support.

Many parents are afraid that children who use calculators will grow up with weak arithmetic skills. Others argue that because calculators are so ubiquitous, children must learn to use these tools as a natural part of their education. Why is the public so divided on this issue?

Calculators in education are a hot topic, but it is important to clear up some confusion. In the public's mind, the question about the use of calculators and computers in education is not "whether" but "when and how." Our studies show the public believes it is vital that students be trained in the use of computers and other new technologies. In *Assignment Incomplete,* a follow-up study to *First Things First,* we found that 80 percent of the public rated these skills as "absolutely essential" in a public school education. Skill in the new technologies has become, in the public's mind, a "fourth R," mastery of which is essential for a young person to make it in today's world of work. Once people feel the more fundamental issues of safety and discipline are being addressed, we anticipate strong public support for teaching children not only how to use calculators but how to achieve full mastery of computers and other technologies.

From the public's perspective, the debate concerns when and how these technologies should be introduced to students, not whether they should learn to master them. The public wants students to learn "old-fashioned" mathematics *before* they get started with calculators and computers. We found that 86 percent of the public believes "kids should memorize the multiplication tables and learn to do mathematics by hand before they go on to use calculators and computers. Otherwise they will never really understand mathematics concepts."

All our research suggests the public sees education in terms of a certain sequential order and a certain set of priorities. People believe that unless the fundamentals are in place, the next steps will be jeopardized. Just as people don't think kids can learn anything in an undisciplined environment, the public doesn't think kids will learn very much advanced math until they acquire the "number facts." What outrages people is the attitude, captured in a statement we heard from a teacher in Connecticut to the effect that "we are living in a high-tech age, why do kids need to know how to make change?" This attitude makes no sense at all to most of the public. Their view is that the shift toward computers and high technology has made basic mathematics even more essential, not less. Their opposition to calculators in school, in other words, is vehement if they believe that calculators will be used as an excuse for not learning traditional math. But the public is equally vehement in its view that a high school graduate must be knowledgeable in the new technologies.

Interestingly, the real division here may not be between teachers and the public, but between teachers themselves and educational theorists. In our study, *Given the Circumstances,* we surveyed over 1,000 teachers. Although associations of mathematics educators usually support the early use of calculators, practicing teachers tend to agree with the public that calculators should only be introduced *after* basic mathematics facts have been mastered. Nearly three-quarters (73 percent) of teachers agree with the public's view that skills in hand calculation should come first. On this issue then, teachers themselves are closer to the public than they are to educational theorists.

Of course, advocates of the early use of calculators say that in the hands of a good teacher, these devices can help young children see the patterns inherent in arithmetic, which in turn will help them learn their multiplication tables. Research on matched classrooms seems to show that by middle school, children who use calculators from an early age are no less able in manipulating basic facts, but are more likely to enjoy mathematics. But I gather from what you have been saying that these types of arguments are not likely to have any effect on parents' attitudes about this issue. Is there anything that might change public opinion?

First, educators should not be surprised if the public is somewhat skeptical of the claims of educational research. Educational experts have a habit of overpromising on the potential of their research findings, only to drop their own recommendations within a few years. Our studies show that teachers are now completely burned out by what we call the "reform du jour" syndrome; many teachers tell us that when a new educational idea comes along, they just go through the motions of supporting it, knowing that the idea will burn itself out before it is ever fully implemented. A little humility on the part of educational experts is thus probably in order.

Having sounded this caution, our advice to educators would be twofold: first, when it comes to calculators and hand calculation, educators should stress "both/and" rather than "either/or." While educators who say hand calculation is unnecessary will be rejected, the public will be more sympathetic to an approach that stresses that both calculators and hand calculation are being taught, and that kids are learning to use calculators in addition to a greater emphasis on hand calculation. Second, the use of calculators and computers should be presented as relevant to preparing students for a high-tech work place (rather than as an alternative to a traditional emphasis on mathematics). The public is, as we have noted, strongly committed to the importance of high-tech education. The same people who are outraged by the idea that traditional mathematics education is being watered down may be pleased that children are getting both their mathematics and a leg up on the new technologies.

First the Basics

One of your own studies shows a big gap between public and expert opinion concerning the need for advanced mathematical, technological, and problem-solving skills: the public believes in the value of basic skills, whereas experts stress higher-order thinking. What accounts for this difference? What can be done to close the gap?

Your question conveys an interpretation of our findings on the public's position that is common, but misleading. You say, "the public believes in the value of basic skills, whereas experts stress higher-order thinking." This way of formulating the problem makes it sound as though the public is content with children learning no more than spelling, hand calculation, and basic English, while experts want children to go way beyond this. The argument, as you pose it, is a dispute between a public calling for "just the basics" and experts emphasizing higher-order thinking.

In fact, what the public is really saying is not "just the basics" but "first the basics." It is not that people think higher-order skills are unimportant, but rather that they are even more concerned about two other issues. The first is the belief that until the basics are acquired, nothing else can be learned at all. Second, they believe that many children in today's schools are not acquiring the basics. The debate about higher-order thinking is, for them, a secondary issue because they are convinced that children will never acquire higher-order skills until they have learned to read, write, count, and behave. Indeed, our focus group respondents sometimes suggest educators' discussion of higher-order skills may, in some cases, be "mumbo jumbo" to disguise the fact that the kids haven't yet learned to read.

Our view is that many national leaders just do not hear what the public is saying on this point. And when educational experts characterize the public's position as a desire to limit education to the basics, they are suggesting that the public's position is a result of nostalgia for the little red schoolhouse and ignorance of the demands of the modern world. But what the public is saying is quite different. In effect, the public is saying to leaders and experts, "the skills they taught in the little red schoolhouse may not be adequate in today's world, but many of our kids aren't even getting those skills. Before you talk to me about higher-order skills, I'd like to hear that our kids can read and write and find this state on a map."

Having said all of that, however, it is true the public is not nearly as sold on the need for higher-order skills as are some of the experts. When people look at others who are successful, they see individuals in the fields of sales, law, management, and a variety of other professions. Even those who started in technical fields, such as engineering, often end up in management. Most of these people do not use skills such as advanced mathematics on a day-to-day basis on the job, or if they do, the skills are not the sort that get taught in the schools anyway. Instead, these skills were learned on the job. It would be interesting to ask a survey question such as "How many people do you know who use calculus in their jobs?"

What people do see all the time, however, is that successful people often have a strong work ethic, common sense, energy, organizational ability, and an ability to relate to many different kinds of people. This perception is even further heightened by the belief that many people with advanced scientific or technical training are out of jobs, and that physics Ph.D.'s are driving taxicabs.

Our research shows, in other words, that most people believe the biggest determinants of success are personal qualities or character. Sixty-four percent of the public says that the factor most important to career success is either "being persistent and having inner drive" (41 percent) or "knowing how to deal with people" (23 percent). Only 27 percent stress an academic preparation as the most important factor.

Interestingly, this is an area in which we found little difference between the views of the public and of educators. In fact, the importance of strong academic preparation receive even lower ratings among teachers themselves than it does among the general public. Only 21 percent of teachers rate an "excellent academic education" as the most important factor for career success. Both the public and educators thus stress personal qualities over academic preparation as a predictor of success in life.

It seems as if the common undercurrent in public opinion about both calculators and basic skills is a belief that learning is sequential—that basic facts must be learned first before moving on to more sophisticated (and possibly more exciting) topics. Isn't that like the old-fashioned approach to piano lessons—two years of scales before playing even a simple tune?

There is no question that the public sees education as a sequential process, and no question that an overemphasis on sequence can kill the joy of learning. The public also has a point, presumably, that too much emphasis on process without any focus on content can also interfere with learning. The public's emphasis on the sequential process may be, in part, a reaction to a perception that educators have leaned too far to the other side. The public does not want education to be boring and stultifying, but it does want content as well as process. If educators are able to show real results, the public will be sympathetic to whatever teaching methods are used. What has happened so far is that educators have sometimes seemed to emphasize methods that are, in the public's judgment, both counterintuitive and ineffective. This combination is, of course, completely fatal in the public's mind.

Education and Employment

Former Labor Secretary Robert Reich argues that most of tomorrow's good jobs will require an advanced technical education, but not necessarily a traditional bachelor's degree. Yet parents pressure schools (and students) to strengthen traditional precollege preparation, holding up calculus as the epitome of high school mathematics. What will it take to convince parents to support a broader vision of high school education—to legitimize strong vocational preparation and encourage their own children to participate?

To respond to this question, we need to look for a moment at attitudes toward higher education generally. We studied this issue in a report entitled *The Closing Gateway*. One of the things that leap out of this research is the conclusion that a college degree has assumed the status a high school diploma had a generation ago. People feel that without a college degree, a young person is unlikely to get a secure and well-paying job. People are convinced, in other words, that a college degree is a necessary step in securing a place in the American middle class. We found that 79 percent of the public feels that high school graduates should go to college "because, in the long run, they will have better job prospects." People are also frightened that, just as a college degree is becoming more essential than ever, it is being priced out of the reach of many families. This phenomenon is contributing in a significant degree to a growing economic anxiety that we are now seeing in every issue we study.

Although the public is more than ever sold on the necessity of a college degree, Americans are less convinced about the value of a college education. Indeed, 77 percent think many young people are "wasting their time and money in college because they don't know what to do with their lives," and 54 percent think it is a problem that "too many people are going to college instead of alternatives where they can learn trades like plumbing or computer repair."

Given these attitudes, we might expect that Secretary Reich's views would resonate with the general public, and that we would see a movement away from higher education and toward vocational programs. Indeed, there is now a growth in "second transfers," in which people with four-year degrees enter community colleges to gain marketable skills, causing debates in some states as to whether people who are coming back for a two-year degree should pay a higher fee for their education.

Nonetheless, the public has not really bought into the argument that college is not the only path to a secure middle-class lifestyle. One recent survey of high school seniors found that 85 percent of the respondents said they planned to get a bachelor's degree (*Chronicle of Higher Education,* May 10, 1996).

What will have to change before people's perceptions and behaviors change? First, our research suggests people are waiting for a clearer signal from employers. Our respondents frequently reported that college-educated young people are displacing older and more experienced workers even in jobs that don't really require a higher education. People see rampant credentialism in the work place, whereby employers give preference to applicants with college degrees. Second, we believe that people are waiting for clearer signals from the schools. In most cases, teachers and guidance counselors are still pushing college and tend to stigmatize vocational education. We did a study of attitudes toward vocational education in Westchester and Putnam Counties in New York. Vocational programs still carry a stigma of being a place for those who cannot cut it academically, and teachers themselves are ambivalent about vocational training.

In other words, let's ask these questions. Do schools offer quality vocational programs, and do they refer some of their strongest students to them? Are well-qualified high school graduates advised to enter postsecondary vocational programs, and do these students get high-paying jobs? When the answers to these questions are consistently positive, I think we will see attitudes toward vocational training shift quickly. The groundwork is there; what people are waiting for are consistent signals.

Let's turn finally to some issues closer to home. What should the public know to understand surveys and polls? Do you find that your research is well understood by reporters and politicians? Is the public too gullible, or perhaps too skeptical, about the merits of survey research?

We find a deep and troubling lack of understanding of survey research, especially among reporters and national leaders. This lack of understanding has contributed to misunderstandings on the part of the public and to a breakdown in communication on many issues. Ironically, the problem has little to do with the mechanics and mathematics of survey research. These aspects are relatively well understood, and most reporters have been trained to ask about things such as sample size and margin of error. But this focus on the technical aspects of surveying has obscured a fundamental problem concerning the meaning of survey research itself.

Dan Yankelovich, who with Cyrus Vance founded Public Agenda, has often drawn attention to a distinction between two kinds of public opinion, which he calls "mass opinion" versus "public judgment." His view is that this distinction is extremely important to the healthy functioning of a complex democracy but poorly grasped by many journalists and politicians. This distinction is, in turn, the foundation for everything we have been trying to do here at Public Agenda for the past 20 years.

The difference between "mass opinion" and "public judgment" can best be explained in terms of an enormous difference in what might be called the quality of public opinion. We describe the public's thinking as mass opinion when people are reacting to issues from a "top-of-the-head" perspective. Mass opinion has a number of characteristics: it is typically ill-informed and unstable, and it can be readily changed either by events or media coverage of them. When the public is in this mode, people can give wildly different answers to survey questions depending on slight variations in how questions are worded. Although they may strongly favor a certain view, their opinions can change dramatically when trade-offs, counter-arguments, or new information are introduced.

Prior to the health care debate in 1992, for example, survey researchers often found nearly universal support for a Canadian-style national health care system. But this support really was an example of mass opinion, and reflected the fact that people wanted a solution to the health care problem but did not know very much about it. Public Agenda found, for example, that although 80 percent of the public supported national health care, only a small fraction defined this kind of coverage as the experts did. As one focus group respondent said, "I'm in favor of national health insurance, but I don't want the government to be involved."

Clearly, mass opinion is not a stable platform for policy reform, because when the public does have an opportunity to debate the issues, national leaders frequently find the consensus is "a mile wide and an inch deep." This discovery often sets up its own reaction, causing another round of public bashing by experts and politicians.

In other areas, however, the public has taken the time to think and work through issues and reach a level of consensus that Yankelovich calls public judgment. People stick with this kind of opinion even when trade-offs are offered, and, in surveys, even when the questions are worded differently. This level of awareness demonstrates solid support for public policy. For example, we recently prepared a study of attitudes toward academic standards for the National Education Summit. The idea of setting clear academic standards for what teachers should teach and students should learn is one that has reached a stable consensus among the public. Our research found that it is supported by 82 percent of the public, with little variation among demographic or geographic groups. Furthermore, people support this idea even when they are faced with possible negative consequences, such as the fact that more children might fail. This is an issue, then, on which the public might be said to have formed a judgment.

The failure to recognize this distinction between mass opinion and public judgment has caused two problems: on the one hand, because national leaders and the media have not been trained to make the distinction, they frequently confuse one type of opinion with the other. Thus they overreact to the public's top-of-the-head responses or ignore issues (such as the demand for standards) on which a stable consensus has been reached.

An even greater problem is that as a society we have not found ways to help the public move from mass opinion to public judgment. Here at Public Agenda we are exploring that very question, but this is, perhaps, a topic for another interview.

I am saying, in other words, that survey research is, as you suggested, poorly understood. And it is tempting to think that the cure for the problem is more education for journalists, the public, and national leaders on the technical aspects of survey research. My view is this would be exactly the wrong response. What we need is not more training in mathematics, but more understanding of psychology and a different philosophy of democracy. Since the days of the great public opinion disasters (such as the mistaken headlines announcing Truman's defeat), public opinion polling has become more and more scientific, and reporters are more thoroughly conversant with how polls are conducted and what they mean. But the art of interpreting public opinion has not kept up with the techniques for surveying public opinion.

Understanding the News

GINA KOLATA
The New York Times

For lack of sufficient quantitative reasoning, both readers and reporters frequently misinterpret medical and scientific news. Overvalued anecdotes, misinterpreted percentages, and unverified studies impede thoughtful analysis of evidence.

Your career as a science writer straddles biology and mathematics. Can you give one or two examples of science stories you have written that require quantitative sophistication beyond what you can assume of your typical reader?

My favorite example of a type of quantitative reasoning that many readers seem unable to understand is why anecdotal evidence is less reliable than data from research such as cohort studies, which in turn are less reliable than data from randomized controlled clinical trials. This distinction was central to the breast implant stories I wrote. Many readers assumed that if women came forward and said that their symptoms—such as fatigue or muscle aches and pains—began after they had implants put in, then that was evidence that the implants caused the symptoms. If the women said they began to feel better after having the devices removed, what more could be asked for in the way of evidence? Of course, such anecdotal evidence can only suggest a cause-and-effect relationship, not demonstrate one.

Anecdotal evidence is also used by alternative medicine practitioners who, for example, sell cancer or AIDS treatments or say that acupuncture cures addictions. I wrote a series of articles on alternative medicine and, in the first story, I spelled out for readers the difference between anecdotal evidence and data from clinical trials.

Another type of quantitative sophistication is required for a story involving charges that trace amounts of chemicals in the environment are causing precipitous declines in sperm counts. The studies alleging declines in sperm count had serious flaws and appreciating those flaws requires quantitative sophistication. For example, one influential study compared sperm counts of donors over the years but failed to ask whether the criteria for being a donor had changed. If 20 years ago donors were selected because they had particularly high sperm counts, a recent decline in sperm counts could mean nothing more than a change in selection criteria.

A third type of quantitative reasoning is necessary to cut through the hype in stories that say, for example, that women are the fastest growing group of AIDS patients, citing a certain percentage increase in incidence compared to last year. The data often look quite different if you look at the raw numbers. After all, a 100 percent increase in disease incidence will occur if you go from one case to two or if you go from 1 million to 2 million, but the significance for public health is very different. The public often looks at stories saying that AIDS in women has increased by a certain percentage and assumes that means there is an epidemic among women, yet the actual numbers tell a very different story.

These are rather basic issues to be confused about. What, if anything, can you assume about the quantitative literacy of your audience? Because you write for the New York Times, *can you assume greater literacy than journalists working for other papers can assume?*

I assume that the audience knows little and does not want to learn any more than necessary to understand the news in a story that is interesting to them for other reasons. I do not think that anyone, myself included, wants to read a newspaper story slowly and carefully because it is dense with new information. And my editors would not publish such a story. The readers of the *New York Times* include scientifically literate people, but also people like my 14-year-old son, and some of my neighbors, many of whom have college degrees but know very little about science and mathematics.

Do you recall any examples of egregious errors made by journalists or public officials due to a fundamental misunderstanding of the mathematical or statistical aspects of some important issue?

Journalists often are completely taken in by anecdotal evidence. A good example is the breast implant story. Journalists were convinced that a few women coming forward with tales of illnesses were proof that the devices caused the illnesses; they simply did not understand the limitations of anecdotal evidence. The result was to fuel the litigation and frighten women. Even now, with solid epidemiology, some journalists still cite anecdotal evidence supplied by plaintiffs' lawyers as though it somehow has equal weight with real science, as though the two types of evidence are equivalent.

Another recent example of journalistic ignorance of quantitative methods was the rush to believe, without evidence from clinical trials, that beta carotene prevents cancer. The studies were tantalizing—some showing that beta carotene levels in the blood were associated with lower risks of cancer and that people who ate foods rich in beta carotene had a lower cancer risk. But I think that few journalists appreciated the limitations of such evidence and few waited for the randomized placebo-controlled clinical trials (which showed that beta carotene supplements do not prevent cancer and may even increase the cancer risk for smokers).

The night that the first study failing to find a beneficial effect from beta carotene came out, my husband was channel surfing and came across Frankie Avalon hawking beta carotene. He was saying that one study found that beta carotene does not work but hundreds say it does, so what are you going to believe—one negative study or hundreds of positive ones? He was, of course, playing to the notion that all evidence is equally convincing and that hundreds of positive studies outweigh one negative one. Those taken in by such arguments do not realize that what matters is how the studies are done, not how many of them there are.

Another example comes from the public schools in Princeton, New Jersey. The lower schools are moving toward classes that combine students from two grades. Many parents are opposed. But the school board was told: "In a review of 64 studies, cited in ASCD's Educational Leadership issue of October 1992, research clearly supports the use of nongraded programs." Sounds impressive, but what, exactly did those 64 studies consist of? Maybe they are of the highest quality, but the point is that the quality of the studies was not addressed. The number was.

Many major national issues—for example, balanced budget, national debt, health insurance, Social Security—rest on complex quantitative arguments such as rates of growth and economic forecasts. In what way does our collective innumeracy impede civic discourse?

It encourages the view, which I deplore, that every opinion is equally valid and that when there is disagreement, the truth lies in the middle. It encourages arguments by sound bite and an increasing gulf between the educated few and the rest of the population.

What strategies do you use in your stories to help readers if they need further quantitative background to understand a story?

I never use jargon, I never assume any knowledge of quantitative reasoning, and I use concrete examples.

Can newspapers provide education in quantitative literacy? Could the media support lifelong learning by enhancing literacy? Or is that inappropriate?

I don't think that is the role of a newspaper. Newspapers present news; they are not in the business of educating, except subtly and by way of explanation. That does not mean articles cannot educate but rather that newspapers are not the forum for an explicit educational effort.

Do you ever get letters from readers who have misunderstood a quantitative aspect of a science story?

All the time. I just got a phone call from a man who misunderstood a story on folic acid fortification. I said that about 2,500 babies are born each year with neural tube defects, a devastating type of birth defect, but that with increased folic acid in the diet, 50 to 80 percent of those cases could be prevented. I also said that approximately 1 old person out of 250,000 has pernicious anemia and that clinical studies done decades ago showed that very high levels of folic acid—nearly three times what people will consume by fortifying food with folic acid—can mask this symptom of the disease, but that it was not clear that any old people would be harmed by the levels of fortification that the F.D.A. is requiring.

The caller asked whether fortification might not do more harm than good because there might be more old people with symptoms masked than babies saved from neural tube defects. First, the caller misunderstood what the data said: that there was a strong likelihood, based on randomized controlled clinical trials, that at least 1,250 babies a year would be saved, as opposed to the vast uncertainty, based on inferences from clinical experience decades ago and not from controlled studies, that any old people would be harmed.

But, perhaps more important, he did not do the simple arithmetic to see whether his interpretation of the data made any sense. If just the lower bound of benefit was realized and 1,250 babies a year were saved, and every one of the old people with pernicious anemia was harmed, there still would be far more babies saved than old people harmed. Suppose the fortification saved 50 percent of the babies, or 1,250 babies, and that it harmed every old person with pernicious anemia. Then, if 1 in 250,000 old people has pernicious anemia and every one of them was harmed, there would have to be 300 million old people in the United States for the number harmed to equal the number of babies saved. But the entire U.S. population is just 250 million!

Most people who read the New York Times *studied some mathematics, at least algebra and geometry, in school, and a good number of your readers use some kind of simple mathematics (e.g., spreadsheets, graphs) in their jobs. Given this, why don't science reporters make greater use of mathemat-*

ics and other quantitative methods as part of their repertoire of strategies for telling their stories?

We try to make our stories easy to read, and most people find equations difficult, even if they did study algebra. Most people do not expect to work when they read a newspaper, and they perceive equations as work even if they can handle them. We do use graphs and charts to illustrate our stories, as a way of emphasizing points we make in the text.

Do reporters suffer from math anxiety? Who is responsible for the low level of quantitative sophistication in journalism?

Some reporters, and some editors, are put off by mathematics. I assume that these reporters and editors steered clear of mathematics courses in school and majored in the humanities. I guess you're asking if journalism schools should have been teaching quantitative reasoning, and I can't answer that because I never went to journalism school. For all I know, it was taught and people forgot what they learned.

For centuries, quantitative reasoning was defined by the third "R." Do we really need more than that today? Aren't arithmetic and simple geometry enough for most people?

I think that quantitative reasoning is one of the most important things people can learn in school. Beyond arithmetic and geometry, students must learn logic, the basics of statistics and data analysis, probability, and the analysis of evidence. An education in quantitative reasoning makes all the difference in the world in people's ability to understand issues of national and personal importance and helps them evaluate in a rational way arguments made by the press, the government, and their fellow citizens.

Quantitative reasoning is becoming even more important as schools rush to embrace the Internet as a source of information for students. Actually, it is a source of disinformation. I have found that it is, if anything, easier to find information on the Internet put out by people with little knowledge but strong opinions, or by charlatans who are promoting or selling things like cancer cures that have no validity, than it is to find scientifically valid data. The problem, of course, is that the trash can be hard to distinguish from the sound information, especially for people with no quantitative reasoning skills.

I think that a much better use of our educational resources would be to provide students with the quantitative reasoning tools needed to evaluate information, rather than to encourage use of the Internet to provide "data" of uneven quality.

Is it reasonable to expect the typical high school graduate to be able to make these distinctions? It sounds as if you need a graduate degree in statistics or information science.

Yes, I think high school graduates can, and should, be able to make such distinctions. A lot can be done without any formal mathematics at all, although I would like to see elementary statistics included in high school curricula. But it is amazing how far kids can go just by learning how to reason quantitatively.

I taught my own children how to reason by discussing news stories with them and explaining how to ask the right questions about data. I never used paper or pencil, a calculator, or a computer. All I did was tell them stories at dinner that involved correct, and incorrect, data analyses. We discussed AIDS statistics and the beta carotene study. We all laughed when Frankie Avalon pointed to his stacks of studies. As a result, my children knew how to reason before they ever reached high school. The problems came when they tried to instruct their friends, which indicates to me that children do not instantly catch on but that, with properly designed instruction using vivid examples, it is possible to produce a quantitatively literate population.

For example, when my son was in seventh grade, he told his friend that randomized controlled clinical trials had failed to demonstrate that sugar makes kids hyperactive. I had explained to him that children whose own parents thought children were made hyperactive when they ate sugar were randomly assigned to drink a sugary lemonade or one sweetened artificially. Neither the children nor those giving out the drinks knew which had the sugar. It turned out that there was no observable difference in the behavior of the children who got sugar and those who did not.

My son's friend listened attentively and then said that was all very fine but, "they didn't try it on me," thereby illustrating that he had missed the point. I think, however, that if that boy had sat in on our dinner conversations for a few years, he would have caught on instantly.

How would you define the level of quantitative reasoning that is appropriate for society to require of all high school graduates?

Every high school graduate should be able to read and understand charts and graphs, but that is only the beginning—like saying they should know how to write a sentence. They also should understand logical arguments and logical fallacies. They should understand the nature of evidence. They should understand such things as the difference between absolute and relative risk. Quantitative literacy, in my view, means knowing how to reason and how to think, and it is all but absent from our curricula today.

People often complain that they hear one thing one day and another thing the next—beta carotene is good for you, then it's bad; fat in the diet causes breast cancer, then it doesn't—and ask how they are supposed to know what to believe. The answer is that they have to learn how to think for themselves, and that is what an education in quantitative reasoning can teach them.

What is the responsibility of newspapers to help people think for themselves? Do newspapers provide enough information to enable readers to make informed judgments?

A good newspaper article will provide enough information, and commentary, to allow readers to think for themselves. But all too often newspapers, even the *New York Times,* do not do that. By neglecting such information, stories can encourage illogical thinking. This is not just a matter of lack of space. Yes, many newspapers keep stories very short, but even long stories can omit crucial discussions that many readers need if they are to think for themselves.

Making Mathematics the Great Equalizer

SHIRLEY MALCOM
Director, Education and Human Resources
American Association for the Advancement of Science

Quantitative literacy is more important than ever, but it depends as much on "unfound" mathematics from everyday life as on textbook mathematics. By fitting the subject to the audience, teachers can use mathematics to reduce rather than enhance differences in students' performance.

Many years ago sociologist Lucy Sells identified mathematics as the "critical filter" for education. Is that metaphor still apt?

I would say that mathematics is even more of a critical filter than ever. If anything, it has ceased being just a gatekeeper for entry into college majors and has become the determinant of participation in courses of study at the high school level such as advanced academic, vocational, and technical programs. Whether your employer is willing to invest in retraining you or upgrading your skills may also be a function of your attitude about and willingness to work at mastering mathematics.

We've now had nearly two decades of supposed education reform since the publication of A Nation at Risk. *Has anything changed?*

I think that a lot of things have changed. More girls than ever are taking mathematics. The charge to take mathematics came from beyond the science, mathematics, and engineering communities. The need for women and minorities to take mathematics has been stressed by higher education and even civil rights groups, who have come to see mastery of mathematics as a civil rights issue. Bob Moses, who at one point put his life on the line to improve the chances of blacks in Mississippi to take control of their political destinies by emphasizing voting rights, is now working to increase the access of black youth to higher-level mathematics for just the reason we are talking about—mathematics is viewed as a critical filter for education and economic empowerment.

You've said on occasion that mathematics doesn't have a "story line" that students can appreciate. What should its story line be?

I would suggest that the relevant story line is that mathematics is everywhere: we should offer students the opportunity to study what they know, what they love, what they care about, what interests them, and then find the mathematics in it. Just think what teachers could do with the crowd estimates at the Million Man March. There's lots of mathematics in that—counting, estimating, probability, geometry—and lots of interest.

Recently, I gave a presentation for the Columbus Ohio Urban Systematic Initiative. The audience was mixed, ranging all the way from students to teachers and principals, from school administrators to people from local industry. A 9-year-old came up and asked me about mathematics because I had said there was mathematics in baseball. We proceeded to have a very long conversation about the different statistics that are used in baseball—won-lost records, earned-run averages, batting averages, etc.—and about all the mathematics that you need to really understand what's happening. Do you really believe I would have held his attention for that long had I not made it meaningful to him?

Mathematics educators are on a campaign to make mathematics inclusive, to serve the many rather than the few. But many parents see mathematics as the ideal vehicle for making distinctions, for sorting the few from the many. How do you see it?

Mathematics has to be made more inclusive for just the reasons that I described earlier. Mathematics is the critical filter for all kinds of things in life: jobs, understanding news and advertising, recognizing information that will protect us from being duped. Unfortunately, people seem really attached to the old nature/nurture controversy. The notion of a math gene still seems to be floating around and we can't seem to get rid of it. No one talks about a reading gene, even though reading is similar in terms of the skills that it represents. Yet we want some way of distinguishing our kids in terms of mathematics ability. I think it's very unhelpful.

Maybe what we have is the "Lake Wobegon" effect—Garrison Keillor's notion of the town where all the children are above average. Maybe this emphasis on a math gene has been one way of affirming our children's specialness (and through them, our own?) Maybe mathematics carries some special "cachet" because of its perceived difficulty.

I think that we should return to an idea espoused by David Hawkins in a 1990 *Daedalus* article: that mathematics and science can in fact be the great equalizers within education. Science and mathematics empower everyone who learns them, regardless of background. Success in these fields depends primarily on diligence and imagination—characteristics that can be found in abundance in all children. We must learn to see in these programs the magnificent opportunity to reduce rather than widen differences through excellent instruction that can have life-changing consequences.

Several years ago, Washington Post *columnist William Raspberry stirred up controversy among mathematics educators by challenging the notion that everyone needs to learn algebra. Just how important is algebra anyway?*

I don't agree with William Raspberry. I think that the same kind of affirming sense that we gain when we have mastered anything challenging applies as well to the study of algebra. In algebra we are not just studying the relationships of numbers to one another but we are also affirming the relationships of numbers to ourselves. We must see algebra as a powerful friend with extraordinary explanatory powers. I believe that mastery removes the mystique that surrounds algebra. Maybe universal algebra mastery would help us get over some of the "math gene" idea.

The notion that we can go from the unknowns to the knowns by identifying relationships, that we can find out what *we* want to know, is a very powerful idea, something of real value that we gain from the study of algebra. But no one sees it because no one has made a serious attempt to demonstrate the relevance of algebra.

Is algebra essential to mathematical literacy? Isn't statistical reasoning more important for today's citizen? What are your priorities for mathematical or quantitative literacy?

I don't know that algebra as a course is essential, but the concepts and ideas embedded in algebra are extremely important, for example, moving from the unknown to the known. Statistical reasoning is also important because it conveys the power to explain. Statistics are used in everyday life and are routinely used to make sense of things that we care about. The explanatory power of statistics is well-illustrated by the way they are used in the popular media. Some aspects of statistics can also be arcane, but those are not the ones that are emphasized to the public. People do understand, because of the media, that statistics are important. But that has not yet been done for algebra, and until it is, we will never gain broad support for "algebra for all."

Many parents bear scars from their own mathematics education, and many children drop mathematics as soon as they can. You've worked a lot with children and families, especially in urban communities. Do you see any way to break the cycle of negativism about studying mathematics?

I believe that the only way the cycle can be broken is to engage parents along with their children in activities where the children can be successful in mathematics, where they can gain power from mathematics, where they can begin to see the importance of mathematics in their lives. Then parents can see that they and their children are capable of handling a lot more of the ideas in mathematics than they think.

You are a scientist. Do you believe that prospective students of science should study a different type of mathematics in high school from that which vocational students study? Can the same curriculum serve all high school students?

I really chafe at the idea that different students need different mathematics. I think we should maintain the same standards and present the same mathematics concepts but using very different examples. We may want to emphasize different topics (going more deeply into statistics, for example, for technical education students who may need these skills in manufacturing settings). Whether that means providing different curricula depends on the attitudes and skills of teachers and their willingness to accommodate individual differences among students.

If schools operate in "lockstep," then there is a problem for everyone. But if students' varying interests are supported and they are encouraged to connect the mathematics that relates to those interests through an appropriate teaching and motivational strategy, then all can be accommodated in the same program.

I do not believe that shop students need less mathematics, however. Studies such as *What Work Requires of Schools* (the SCANS report) and *America's Choice: High Skills or Low Wages* (the report of the Commission on the Skills of the American Workforce) emphasize that curricular ghettos will not work in the new economic reality. Mathematics needs to make explicit connections with other subject areas, with the world of work, and with people's everyday lives.

What about honors courses and special programs for the gifted and talented? Don't these children deserve courses appropriate to their abilities and interests?

Let me tell you about an experience I had just a few months ago. I was visiting a high school in a university town where the science teachers had decided on their own that there was really no difference between honors and regular precollege courses. The way they were taught, they were really the same course. In the teachers' view, tracking was not compatible with where they wanted to go with standards.

Well, the community, especially the university faculty, was up in arms. They perceived this as a move to endorse watered-down science. And besides, they said, not all these kids are equally motivated, not all of them are this or that, etc. They couldn't possibly imagine their children being in the same classes with the others. It was really heavy. People were not speaking with each other, and I got e-mail asking if I really understood what I was getting into.

You wouldn't believe the posturing that I encountered. University faculty felt they had to establish their credentials as faculty before they would enter the dialogue. They had looked at analyses by educational researchers J. A. and C. C. Kulik and others that found differences in student performance in tracked systems, so they came prepared to challenge the position that I was taking—that our landscape has changed out from under us.

It really took people by surprise when I didn't come at the issue from the perspective of tracking. Instead, I asked, what things do we want all children to know and be able to do? Where are the standards taking us? Most people did agree that slicing the student population into horizontal bands is not particularly helpful if your goal is to move toward those standards.

But as you know, mathematics is the most tracked of all subjects, and most mathematics teachers believe strongly that tracking leads to the most effective education for all students. Does tracking in mathematics classes enhance or diminish the prospects for a quantitatively literate population?

The issues in mathematics are particularly problematic. Tracking supports mathematics education as we have known it—topic-specific, disconnected from rich examples, a race through the concepts and the textbook. Let's face it, this stuff is just a lot easier for some people to master than for others. I do believe there are people who learn mathematics more easily than others, perhaps because they have "hard wiring" that enables them to do mathematics the way it is currently taught. But that is not the same as having a math gene.

Teachers seem to allow very little variation in the way mathematics is taught. If you made real changes, you'd find interesting surprises, not only about people's capacity to learn but also about mathematics itself. You'd discover lots of "unfound" mathematics that exists in everyday life, and in people's minds, but not in the textbook. We've barely begun to examine the deep implications of how we must rethink the discipline and the way that it is taught in order to be inclusive. The linear topical approach may not support the way most people really operate.

Many problems are due to the lack of diversity in the way mathematics is taught and accessed. People who need different routes through the material may only "find" mathematics if they continue to demand it and if they insist that they have a right to it! Perhaps mathematics and quantitative literacy would be advanced if we mixed these people together and offered them complex, real-world problems and many alternative ways of addressing them.

This idea of turning content on its head to fit the subject to the audience is the real uncharted territory of mathematics education. I can do this kind of analysis in the biological sciences because that is an area I have studied extensively. But I do not know mathematics well enough to make a similar analysis. It will take people with deep understanding of content and a willingness to perceive the possible advantages of breaking out of the box of content and the linear approach. Until this happens, our own academic community will be our worst enemy in this discussion.

It is not just education, but the entire scientific enterprise that is at stake here. If we cannot show how our fields of study are important to the typical man and woman, we cannot expect support. This goes beyond equity, beyond any kind of special pleading. It goes to the core of our knowledge base and how it relates to the real world. That's why we need people in the mathematics community who can find ways to address these difficult and important issues.

Numeracy: Imperatives of a Forgotten Goal

IDDO GAL
University of Haifa

There are no "word problems" in real life. Adults face quantitative tasks in multiple situations whose contexts require seamless integration of numeracy and literacy skills. Such integration is rarely dealt with in school curricula.

A key goal of school is preparing students for life as adults. However, when it comes to quantitative literacy, the form and content of this preparation are not at all clear. Since numeracy should serve as the primary (but not exclusive) basis for determining the scope, content, and methods of mathematics instruction and assessment at the K–12 level, it is especially critical to define and understand numeracy.

Some years ago, the Numeracy Project at the National Center on Adult Literacy began to address this issue. Other efforts to define aspects of the mathematical requirements of adult life can be found in publications representing the perspectives of different players in the educational and policy-making arenas, such as:

- *Employers:* For example, see *Workplace Basics,* published by a task force of the American Society of Training and Development[1] and the report by the Secretary of Labor's Commission on Achieving Necessary Skills (SCANS).[2]
- *The K–12 education community:* For example, see *Being Numerate*[3] and studies by the National Council of Teachers of Mathematics (NCTM)[4] and the American Association for the Advancement of Science,[5] as well as the Cockcroft report from the United Kingdom.[6]
- *The adult education community:* For example, see the recommendations of teachers of adults (e.g., in literacy, the GED, and work place

programs), the most recent by the Adult Numeracy Practitioners Network in the United States.[7]

- *Large-scale assessment projects:* For example, see surveys such as *Adult Literacy in America*[8] and *Adult Literacy in Canada*[9] and efforts to test the functional skills of adult learners, such as the *GAIN Appraisal Program.*[10]

It is clear from these and related documents that there is no agreement concerning the goal of school education in mathematics. In developing their recommendations, different associations and projects employ different "lenses" and make different assumptions about the contexts of people's lives. These sources, supplemented by the results of interviews with adults and adult educators, suggest that there are multiple life contexts in which adults may need to deal with situations involving numbers, quantities, mathematical concepts, and algorithmic processes,[11] including:

- *Home:* Shopping, home repairs, cooking, coordinating schedules, understanding prescription labels;
- *Recreation:* Planning a trip or party, designing a crafts project, knitting;
- *Personal finance:* Filling out tax forms, monitoring expenses, paying bills, negotiating a car loan, planning for retirement;
- *Informed citizenship:* Comprehending a poll discussed on TV or crime figures reported in a local newspaper;
- *Social action:* Participating in fund-raising activities or helping to design a survey for a local action group, debating the environmental implications of a proposed construction project;
- *Work place:* Shipping merchandise, measuring quantities, computing amounts of materials needed, reading assembly instructions, retrieving data from a computer system, learning statistical process control;
- *Passing tests:* Taking a college entrance exam or a technical certification test;
- *Further education:* Taking college-level courses or enrolling in post-secondary technical training; and
- *Active parenting:* Helping children with mathematics homework, understanding scores on standardized tests and statistics about schools.

Reaching Goals

Results from the recent National Adult Literacy Survey[12] showed that relatively few U.S. students graduate from school equipped to handle these diverse quantitative tasks. Roughly 50 percent of the people interviewed for this large-scale study, including many adults with high school or post-secondary credentials, had major difficulty with many of these functional operations. To ensure realism, the tasks emphasized in this survey simulated diverse real-world situations, requiring that necessary data be extracted from different forms or documents, that computational operations be inferred from printed directions, and that quantitative arguments be comprehended although embedded in technical documents or newspaper prose. Similar results were also found in other industrialized countries.[13]

Further, findings from the National Assessment of Educational Progress (NAEP) have shown repeatedly over the last 25 years that many high school graduates leave school without adequate numeracy skills. Mullis, Dossey, Owen, and Phillips[14] reported, for instance, that only 46 percent of the high school seniors tested as part of the 1990 NAEP assessment demonstrated consistent mastery of fractions, decimals, percentages, and simple algebra (topics that they should have mastered much earlier). Performance was lower among Hispanic and African-American students. Moreover, many of the weakest students (as many as one in four) were not included in this study since they were no longer in school.

A growing research literature suggests that many students have trouble transferring and applying mathematical skills to new contexts, in or out of the classroom.[15] These findings corroborate informal observations reported by many mathematics teachers: too many students have trouble dealing with presumably standard problems that they just "successfully" practiced, once minor elements of the problems are changed.

U.S. high schools have historically stressed, explicitly or implicitly, developing students' ability to a stage where they can handle abstract, college-related subjects such as advanced algebra and calculus. Yet fewer than 50 percent of all students who graduate from high school enter college, and many of these take no further mathematics in college. Students who do not cope well with abstract topics—which many high school mathematics teachers view as "real" mathematics—are usually banished into "general mathematics" or "consumer mathematics" courses that many educators consider a dead end[16] and mathematically uninteresting. Such students receive decidedly insufficient attention from mathematics educators.

Several conjectures can be advanced to explain why school mathematics focuses on the abstract and "academic" and gives relatively little attention to learning skills. One very likely explanation is that mathematics educators overemphasize "internal" views of the goals of mathematics education and do not sufficiently balance them against "external" views.

The internal view is presently in transition. The old view that mathematics instruction should aim to teach number facts and computational procedures is being gradually replaced (or supplemented) by a "new" view—

advanced through the bold efforts of organizations such as NCTM—that emphasizes problem solving, reasoning, and communication; the influences of the revolution in cognitive science;[17] and constructivist ideas in education. This new paradigm is sometimes justified by claims that students should spend more time learning what mathematicians *do* (e.g., conjecture, experiment, check hypotheses, verify results) than what mathematicians *know* (i.e., computational rules, formulas, proofs).

While this new internal view offers exciting possibilities for improving learning and achievement in mathematics, it still pays insufficient attention to the external view, which emphasizes what people (rather than mathematicians) need to be able to do and how people actually function beyond the school walls.

Becoming Numerate

The term "numeracy" focuses educators at all levels on a primary goal that applies to all learners, becoming numerate, before they contemplate the means and processes of achieving this goal, i.e., mathematics education. To be sure, the term numeracy, like literacy, has multiple definitions, ranging from those that emphasize basic computational skills to those that encompass a broad and quite advanced range of skills and dispositions.[18] The important goal of developing "mathematical power," introduced by NCTM,[19] encompasses several but not all aspects of the "high-end" conceptions of numeracy.

Being numerate is much more than knowing mathematics, and numeracy is not the same as mathematics. We use numeracy to describe an aggregation of skills, knowledge, beliefs, dispositions, habits of mind, communication capabilities, and problem-solving skills that people need to autonomously engage in and effectively manage situations in life and at work that involve numbers, quantitative or quantifiable information, or textual information that is based on or has embedded in it some mathematical elements.

Most mathematics problems that students work on are contrived or are presented out of context. In contrast, real-life numeracy situations are always embedded in a context that has some personal meaning to the people involved. Such situations are rarely emphasized in school-based mathematics education.

Real problems can be thought of as scattered inside a "numeracy task space," a space defined by dimensions such as the nature of the required response and the *literacy* demands involved. These situations can be quite diverse, as the following examples illustrate:

- *Computational situations:* These demand the computation of a single number by applying one or more of the four basic operations or other simple arithmetic procedures to numbers or quantities clearly specified in the situation. The figure computed is always "right" or "wrong," regardless of how it was reached. An example is calculating the total price of the products purchased when shopping, or finding the area of a room that has to be carpeted or painted (before the materials are ordered).
- *Decision-making situations:* These demand that people gather and evaluate multiple pieces of information to determine a reasonable or optimal course of action, typically in the presence of conflicting goals and constraints or under conditions of uncertainty. Common examples include tasks requiring handling and optimizing resources such as money, supplies, or time.[20] Other examples are tasks involving choice,[21] such as when people have to decide which of several apartments to rent or which loan schedule is the most manageable for them.
- *Interpretive situations:* These demand that people make sense of, and grasp the implications of, verbal or text-based messages that may rest on quantitative issues but that do not involve direct manipulation of numbers; for example, when reading a newspaper or watching news on TV, being faced with a report on the results of a recent poll or a medical experiment, possibly involving references to percentages, averages, rate changes, etc. Such situations may sometimes involve no numbers at all but still refer to important ideas that are part of mathematics or statistics, as when samples, bias, correlation, or causality are implied.

It is not uncommon for the literacy demands of numeracy situations to be more constraining than the mathematical demands. When adults have trouble dealing with numeracy situations (e.g., choosing the better of two medical benefits packages, a task used in the National Adult Literacy Survey),[22] the problem is likely not a lack of mathematical understanding alone, but as much or more a lack of sufficient "document literacy," poor critical thinking, and inadequate reading comprehension skills. Thus, teaching students mathematics on the basis of the internal view of the subject may not prepare them to handle effectively those many numeracy situations that also require literacy skills. This is especially true of interpretive situations, which are seldom represented well in mathematics textbooks.

Most school mathematics problems have "right" answers and solutions that can be verified by the teacher. Yet most numeracy situations do not have solutions that can be classified as right or wrong. In fact, adults do not "solve" situations: there are no word problems in real life. Rather, adults *manage* situations. When faced with real-life numeracy situations, adults can decide on one of several courses of action, based on an assessment of their personal goals and the situational demands, the severity of the consequences, and personal and situational resources. They can ask, for example, "What am I trying to achieve?" "How accurate should I be?" "Can I afford to make mistakes?" "How much time or energy do I have?" "What technical aids (e.g., calculators) can I use?" "Should I get any help?"

Based on such considerations, adults will plan, choose, execute, monitor, or revise a course of action. This means that even in a situation that a mathematics teacher would classify as "computational," people may find it fitting to sacrifice the precision or quality of their response (for example, by estimating a tip rather than calculating it exactly), or aim for a safety margin (e.g., by overestimating the amount of food to order for a company dinner). The response may be reached in a computationally inefficient way, but still be reasonable within the demand characteristics of the situation (and perhaps even demonstrate considerable mathematical understanding!). Another example is the "satisficing" strategy[23] that adults often use to manage decision-making situations: in order to reduce mental and emotional load, they make decisions without first considering all the possible consequences of each course of action (e.g., not checking out all car dealerships in the city when looking for a good bargain).

Many adults—including those who are highly educated—frequently elect to avoid a task with quantitative elements, or address only a portion of it "to make life easier," or delegate responsibility by asking a family member or a salesperson for help. Such actions may result from people's self-perception of not being "good with numbers,"[24] a phenomenon that can be better attributed to "math abuse" and poor teaching practices during prior schooling than to people's true ability.

Implications

The fact that many numeracy tasks require adults to integrate seamlessly both numeracy and literacy skills has mostly gone unnoticed by the mathematics education community. The instructional implications of this have not been explored, even though this integration of skills is critical to enable adults to be smart consumers and informed citizens. The introduction of the important "communication standard" by NCTM[25] has helped a lot in this regard, yet mathematics teachers may still be more willing to adopt new communication tasks (e.g., student journals, project logs) that support the internal view than tasks that are based on real-world numeracy situations.

Many mathematics educators argue that learning the abstract principles and ideas underlying advanced mathematics topics is sufficient for students to be able to manage real-world numeracy situations. Although this may indeed hold true for some students, the range of situations—and hence the range of skills and dispositions required for effective management of those situations in work, daily life, and civic contexts—is wider and different in important respects from what has traditionally been addressed by the K–12 mathematics curriculum.

To effectively educate all students, mathematics educators should take into account what has been termed here the "external view" when setting goals and planning instructional experiences. This suggestion is supported by growing awareness of the educational benefits of embedding learning in meaningful contexts[26] and of connecting knowledge to its applications.[27] Seen in this way, the goal of preparing *all* students to cope effectively with numeracy situations as adults will not water down the mathematics curriculum but can equally serve college-bound and non-college-bound students.

As Foerch[28] reminds us, we teach *people*, not mathematics. Thus the essential goal of mathematics education should be not to teach mathematics, but to help students become numerate—not only to acquire a solid foundation in the theory and processes of mathematics and statistics but also to learn to manage numeracy situations effectively and to fully realize the implications of poor management of such situations (e.g., loss of money, wasted resources, poor choices).

Moreover, to autonomously engage numeracy situations in a dynamic, fast-changing, and information-laden world, students should also be comfortable enough with mathematics and its applications (what Cockcroft[29] calls "at-homeness with numbers") so they will later be willing to invest in further mathematics-based learning, formal or informal, when life so demands. Unless numeracy becomes a key goal of mathematics education, and unless educators ameliorate rather than exacerbate students' "math fright," only some of those presently being served by the mathematics education community will become fully numerate in years to come. But some is not good enough; all students need to become numerate.

Endnotes

1. A. P. Carnevale, L. J. Gainer, and A. S. Meltzer, *Workplace Basics: The Essential Skills Employers Want* (San Francisco: Jossey-Bass, 1990).
2. Secretaries Commission on Achieving Necessary Skills, *What Work Requires of Schools* (Washington, D.C.: U.S. Department of Labor, 1991).

3. S. Willis, ed., *Being Numerate: What Counts?* (Melbourne, Australia: Australian Council for Educational Research, 1990).
4. National Council of Teachers of Mathematics, *Curriculum and Evaluation Standards for School Mathematics* (Reston, Va.: National Council of Teachers of Mathematics, 1989).
5. American Association for the Advancement of Science (Project 2061), *Benchmarks for Science Literacy* (New York: Oxford University Press, 1993).
6. W. H. Cockcroft, *Mathematics Counts* (London: Her Majesty's Stationery Office, 1982).
7. D. Curry, M. J. Schmitt, and W. Waldron, *A Framework for Adult Numeracy Standards: The Mathematical Skills and Abilities Adults Need to Be Equipped for the Future,* Final Report from the System Reform Planning Project of the Adult Numeracy Practitioners Network (Washington, D.C.: National Institute for Literacy, 1996).
8. I. S. Kirsch, A. Jungeblut, L. Jenkins, and A. Kolstad, *Adult Literacy in America: A First Look at the Results of the National Adult Literacy Survey* (Washington, D.C.: National Center for Education Statistics, U.S. Department of Education, 1993).
9. Statistics Canada, *Adult Literacy in Canada: Results of a National Study* (Ottawa, Canada: Ministry of Industry, Science and Technology, 1991).
10. J. Simon, C. John, and P. Rickard, *GAIN Appraisal Program* (San Diego, Calif.: Comprehensive Adult Student Assessment System, California State Department for Social Services, 1990).
11. I. Gal, *Issues and Challenges in Adult Numeracy,* Technical Report TR93–15 (Philadelphia: University of Pennsylvania, National Center on Adult Literacy, 1993).
12. *Adult Literacy in America.*
13. R. Wickert, *No Single Measure: A Survey of Australian Adult Literacy. Summary Report* (Canberra, Australia: Department of Employment, Education and Training, 1989); *Adult Literacy in Canada.*
14. I. V. S. Mullis, J. A. Dossey, E. H. Owen, and G. W. Phillips, *The State of Mathematics Achievement: NAEP's 1990 Assessment of the Nation and the Trial Assessment of the States* (Washington, D.C.: U.S. Department of Education, National Center for Education Statistics, 1991).
15. T. N. Carraher, A. D. Schliemann, and D. W. Carraher, "Mathematical Concepts in Everyday Life," in *Children's Mathematics,* G. B. Saxe and M. Gearhart, eds. (San Francisco: Jossey-Bass, 1988), 71–88; L. Ginsburg and I. Gal, "Linking Informal Knowledge and Formal Skills: The Case of Percents," in *Proceedings of the 17th Annual Meeting of the North American Chapter of the International Group for Psychology of Mathematics Education,* D. T. Owens, M. K. Reed, and G. M. Millsaps, eds. (Columbus, Ohio: ERIC Clearinghouse for Science, Mathematics, and Environmental Education, 1995); M. K. Singley and J. R. Anderson, *The Transfer of Cognitive Skill* (Cambridge, Mass.: Harvard University Press, 1989).
16. L. A. Steen, "Does Everybody Need to Study Algebra?" *Mathematics Teacher* 85:4 (1992) 258-260.
17. J. Bruer, *Schools for Thought: A Science of Learning in the Classroom* (Cambridge, Mass.: Massachusetts Institute of Technology Press, 1993).
18. D. Baker and B. Street, "Literacy and Numeracy: Concepts and Definitions," in *Encyclopedia of Education,* T. Husen & E. A. Postlethwaite, eds. (Pergamon, 1994).
19. *Curriculum and Evaluation Standards for School Mathematics.*
20. *What Work Requires of Schools.*
21. R. Clemen and R. Gregory, "Preparing Adult Students to Be Better Decision Makers," in *Numeracy Development: A Guide for Adult Educators,* I. Gal, ed. (Cresskill, N.J.: Hampton Press (forthcoming)).
22. *Adult Literacy in America.*
23. A. Newell and H. Simon, *Human Problem Solving* (Englewood Cliffs, N.J.: Prentice-Hall, 1972).

24. S. Tobias, *Overcoming Math Anxiety* (New York: Norton, 1993).
25. *Curriculum and Evaluation Standards for School Mathematics.*
26. J. S. Brown, A. Collins, and P. Duguid, "Situated Cognition and the Culture of Learning," *Educational Researcher* 18:1 (1989) 32-42.
27. *What Work Requires of Schools.*
28. J. Foerch, "Characteristics of Adult Learners of Mathematics," in *Numeracy Development: A Guide for Adult Educators,* I. Gal, ed. (Cresskill, N.J.: Hampton Press (forthcoming)).
29. *Mathematics Counts.*

National Indicators of Quantitative Literacy

JOHN A. DOSSEY
Illinois State University

Literacy surveys now include both "document" and "quantitative" scales. U.S. students and adults continue to perform poorly, however, both in terms of internal expectations and international comparisons.

The quest for literacy, particularly quantitative literacy, has been a hallmark of many of the national recommendations for education reform and improvement over the past decade. Documents from *A Nation at Risk*[1] to the Secretary of Labor's Commission on Achieving Necessary Skills report, *Learning a Living*,[2] showcase the importance of mathematical knowledge and skills across the full range of preparation for a productive life.

Historically, the status of quantitative literacy in the United States has been monitored primarily in terms of the productivity of precollege programs in mathematics. Indeed, the chief barometer of quantitative literacy has often been the annual release of SAT scores. However, with the emergence of standards for disciplinary subjects in U.S. schools and the analysis of U.S. students' performance on both national and international examinations in mathematics and the sciences,[3] new emphasis is focused on the importance of mathematics for *all* students. This reawakening has necessitated a careful inspection of the nature of quantitative literacy.

Quantitative literacy might be viewed as the level of mathematical knowledge and skills required of all citizens. The National Council of Teachers of Mathematics' *Curriculum and Evaluation Standards for School Mathematics*[4] took the stance that school mathematics needed a reshaping that focused its power on achieving a core set of understandings, skills, and beliefs for all students. This view was echoed in a recent article contrasting the mathematics required for work and life and the mathematics required for school courses.[5]

Document Literacy and Quantitative Literacy

Given this push for higher levels of quantitative literacy, let us examine what this term really means. Clearly, not all citizens need a command of calculus and statistics to be productive members of society. However, they do need a solid intuition relating to many key concepts from the world of measurement and data that underlie these specialized bodies of knowledge. Further, they need to know how to both compute and estimate in situations requiring numerical analyses.[6]

Starting in 1986, the National Assessment of Educational Progress (NAEP) attempted to anchor a study of literacy among U.S. adults with an assessment of the skills of young adults (age 21 to 25) on tasks related to three literacy scales: prose literacy, document literacy, and quantitative literacy. These scales measure the ability to apply the knowledge and skills required to understand and use information from texts similar to newspapers and popular magazines; to locate and use information contained in job applications, schedules, maps, tables, and indexes (e.g., the S&P 500); and to apply arithmetic operations embedded in printed materials, such as recording checks in a check register, computing the amount of a tip, completing an order form, or determining the amount of a loan.[7] The latter two areas clearly fall within what most people would call quantitative literacy.

The results of this first attempt by NAEP to measure adult literacy led to the conclusion that while illiteracy was not a major national problem, neither could we say that everyone was literate. Even those most recently in school—those in their early twenties—fell in the middle on the scales. Clearly, such performance was less than what the nation required in human talent for long-term international competitiveness.[8]

The NAEP results concerning document literacy showed that tasks became more difficult as the categories of information in the documents increased, as the amount of distracting irrelevant information increased, and as the central question was less clearly linked to the information needed to answer it. Eighty-four percent of respondents were able to handle tasks such as reading a paycheck stub, but only about 60 percent could handle a multiple-choice question using information from stacked bar graphs. More shocking was the finding that only about one in five could respond appropriately to the bus schedule problem shown in Figure 1.[9]

In the area referred to as quantitative literacy, 72 percent of those tested could handle tasks such as totaling a deposit slip and correctly subtracting on the check register, but only about 40 percent could correctly figure the change they should get back from a meal and the amount of tip they should leave, given what they had ordered, the amount of money they had tendered, and a copy of the menu. To say the least, this finding also

ROUTE 5

VISTA GRANDE

This bus line operates Monday through Saturday providing "local" service to most neighborhoods in the northeast section

Buses run thirty minutes apart during the morning and afternoon rush hours Monday through Friday

Buses run one hour apart at all other times of day and Saturday

No Sunday, holiday or night service.

You can transfer from this bus to another headed anywhere else in the city bus system

	OUTBOUND from Terminal						INBOUND toward Terminal					
	Leave Downtown Terminal	Leave Hancock and Buena Ventura	Leave Citadel	Leave Rustic Hills	Leave North Carefree and Oro Blanco	Arrive Flintridge and Academy	Leave Flintridge and Academy	Leave North Carefree and Oro Blanco	Leave Rustic Hills	Leave Citadel	Leave Hancock and Buena Ventura	Arrive Downtown Terminal
							6:15	6:27	6:42	6:47	6:57	7:15
							6:45	**6:57**	**7:12**	**7:17**	**7:27**	**7:45** Monday through Friday only
	6:20	6:35	6:45	6:50	7:03	7:15	7:15	7:27	7:42	7:47	7:57	8:15
	6:50	**7:05**	**7:15**	**7:20**	**7:33**	**7:45**	**7:45**	**7:57**	**8:12**	**8:17**	**8:27**	**8:45** Monday through Friday only
	7:20	7:35	7:45	7:50	8:03	8:15	8:15	8:27	8:42	8:47	8:57	9:15
AM	**7:50**	**8:05**	**8:15**	**8:20**	**8:33**	**8:45**	**8:45**	**8:57**	**9:12**	**9:17**	**9:27**	**9:45** Monday through Friday only
	8:20	8:35	8:45	8:50	9:03	9:15	9:15	9:27	9:42	9:47	9:57	10:15
	8:50	**9:05**	**9:15**	**9:20**	**9:33**	**9:45**	**9:45**	**9:57**	**10:12**	**10:17**	**10:27**	**10:45** Monday through Friday only
	9:20	9:35	9:45	9:50	10:03	10:15	10:15	10:27	10:42	10:47	10:57	11:15
	10:20	10:35	10:45	10:50	11:03	11:15	11:15	11:27	11:42	11:47	11:57	12:15
	11:20	11:35	11:45	11:50	12:03	12:15	12:15	12:27	12:42 p.m.	12:47 p.m.	12:57 p.m.	1:15 p.m.
	12:20	12:35	12:45	12:50	1:03	1:15	1:15	1:27	1:42	1:47	1:57	2:15
	1:20	1:35	1:45	1:50	2:03	2:15	2:15	2:27	2:42	2:47	2:57	3:15
	2:20	2:35	2:45	2:50	3:03	3:15	3:15	3:27	3:42	3:47	3:57	4:15
	2:50	**3:05**	**3:15**	**3:20**	**3:33**	**3:45**	**3:45**	**3:57**	**4:12**	**4:17**	**4:27**	**4:45** Monday through Friday only
PM	3:20	3:35	3:45	3:50	4:03	4:15	4:15	4:27	4:42	4:47	4:57	5:15
	3:50	**4:05**	**4:15**	**4:20**	**4:33**	**4:45**	**4:45**	**4:57**	**5:12**	**5:17**	**5:27**	**5:45** Monday through Friday only
	4:20	4:35	4:45	4:50	5:03	5:15	5:15	5:27	5:42	5:47	5:57	6:15
	4:50	**5:05**	**5:15**	**5:20**	**5:33**	**5:45**	**5:45**	**5:57**	**6:12**	**6:17**	**6:27**	**6:45** Monday through Friday only
	5:20	5:35	5:45	5:50	6:03	6:15						
	5:50	**6:05**	**6:15**	**6:20**	**6:33**	**6:45**						Monday through Friday only
	6:20	6:35	6:45	6:50	7:03	7:15						

To be sure of a smooth transfer tell the driver of this bus the name of the second bus you need

On Saturday afternoon, if you miss the 2:35 bus leaving Hancock and Buena Ventura going to Flintridge and Academy, how long will you have to wait for the next bus?

Figure 1.
Level IV document literacy item from the National Survey of Adult Literacy (NALS).

came as a shock! Clearly, this first attempt to measure adult literacy showed that young adults lack many of the "walking around" quantitative skills we would expect of almost any citizen.

Unlike the more frequently administered NAEP assessments in mathematics[10] reported for students in grades 4, 8, and 12, the assessment of adult literacy measured a full cross section of young adults, those completing school and those not. The first NAEP assessment in mathematics also made an attempt to evaluate both 17-year-olds in school and not, as well as adults age 26 to 35. The results of this assessment showed, in general, a slight increase in performance on social and consumer questions for adults, but a decrease in knowledge of technical mathematical terms as the subjects increased in age.[11]

Based on these early attempts to measure aspects of quantitative literacy, others began to consider how we might define literacy broadly in terms that relate specifically to the mathematics of school, the demands of employment, and the needs of society. "Literate" is defined in the *American Heritage Dictionary of the English Language* first as "able to read or write" and second as "knowlegeable; educated." These definitions provide some direction for understanding the meaning of quantitative literacy.

Clearly, everyone who is quantitatively literate would be capable of using written, spoken, or graphic sources dealing with number, spatial, or data information in achieving goals and functioning in everyday life. In short, quantitatively literate people are capable of manipulating aspects of mathematical knowledge to understand, predict, and control situations important to their lives. Such people have the ability to reason in numerical, data, spatial, and chance settings; to integrate and apply mathematical concepts and procedural skills; and to develop and interpret models related to the problems they encounter.

National Survey of Adult Literacy

The 1992 National Adult Literacy Survey used a definition of literacy similar to that employed in the NAEP. This assessment reported the data in terms of levels of literacy rather than just as a score on a 0 to 500 scale as in the 1986 NAEP assessment. The NALS assessment defined literacy as "Using printed and written information to function in society, to achieve one's goals, and to develop one's knowledge and potential."[12] This paralleled the definition of literacy stated in the 1991 National Literacy Act, to wit: "An individual's ability to read, write, and speak in English and compute and solve problems at levels of proficiency necessary to function on the job and in society, to achieve one's goals, and to develop one's knowledge and potential."[13]

The NALS effort adopted the three forms of literacy—prose, document, and quantitative—used in the 1986 NAEP assessment, distinguishing data-handling and graphic skills from those more directly related to computation and numerals. The 0 to 500 scale was divided into five regions, and anchoring methods were used to relate performance on a given item to one of the five regions along the scale. Items anchored at various points were then used to characterize the given levels. Table 1 illustrates the levels of document and quantitative literacy that resulted from this analysis.[14]

The sample questions shown in Figures 1 to 4 offer further insight into the levels of performance evaluated in the NALS assessment. The bus schedule item in Figure 1 was classified as a Level IV item on the NAEP document literacy scale and was one of the items also used in NALS. The table of information in Figure 2 illustrates a Level V document literacy task. NALS participants were asked to write a brief paragraph summarizing the extent to

Table 1.
NALS Levels of Document and Quantitative Literacy

Document Literacy	*Quantitative Literacy*
1. Tasks at this level tend to require the reader either to locate a piece of information based on a literal match or to enter information from personal knowledge onto a document. Little, if any distracting information is present.	1. Tasks at this level require the reader to perform single, relatively simple arithmetic operations, such as addition. The numbers to be used are provided and the arithmetic operation to be performed is specified.
II. Tasks at this level are more varied than at Level I. Some require the reader to match a single piece of information; however, several distracters may be present, or the match may require low-level inferences. Tasks at this level may also ask the reader to cycle through information in a document or to integrate information from various parts of a document.	II. Tasks at this level typically require the reader to perform a single operation using numbers that are either stated in the task or easily located in the material. The operation to be performed may be stated in the question or is easily determined from the format of the material (for example, an order form).
III. Some tasks at this level require the reader to integrate multiple pieces of information from one or more documents. Others ask the reader to cycle through rather complex tables or graphs that contain information that is irrelevant or inappropriate to the task.	III. In tasks at this level, two or more numbers are typically needed to solve the problem, and these must be found in the material. The operation(s) needed can be determined from the arithmetic relation terms used in the question or directive.
IV. Tasks at this level, like those at the previous levels, ask the reader to perform multiple-feature matches, cycle through documents, and integrate information; however, they may require the reader to provide numerous responses but do not designate how many responses are needed. Conditional information is also presented in the document tasks at this level and must be taken into account by the reader.	IV. These tasks tend to require the reader to perform two or more sequential operations or a single operation in which the quantities are found in different types of displays, or the operations must be inferred from semantic information given or drawn from prior knowledge.
V. Tasks at this level require the reader to search through complex displays that contain multiple distracters, to make high-level test-based inferences, and to use specialized knowledge.	V. These tasks require the reader to perform multiple operations sequentially. They must disembed the features of the problem from text or rely on background knowledge to determine the quantities or operations needed.

which parents and teachers agreed or disagreed on the statements about issues pertaining to parental involvement at their school.[15]

Figures 3 and 4 provide examples of tasks at Levels IV and V from the quantitative literacy section of NALS. The item in Figure 3 asked participants to make a quick-comparison shopping estimate concerning the price-

Parents and Teachers Evaluate Parental Involvement at Their School

Do you agree or disagree that . . . ?

	Total	Level of School		
		Elementary	Junior High	High School
		percent agreeing		
Our school does a good job of encouraging parental involvement in sports, arts, and other nonsubject areas				
Parents	77	76	74	79
Teachers	77	73	77	85
Our school does a good job of encouraging parental involvement in educational areas				
Parents	73	82	71	64
Teachers	80	84	78	70
Our school only contacts parents when there is a problem with their child				
Parents	55	46	62	63
Teachers	23	18	22	33
Our school does not give parents the opportunity for any meaningful roles				
Parents	22	18	22	28
Teachers	8	8	12	7

Source: The Metropolitan Life Survey of the American Teacher, 1987

Figure 2.
Level V document literacy item from NALS.

per-ounce of creamy peanut butter. To answer, a respondent had to locate the appropriate data, distinguishing between the two types of peanut butter information presented. After that, the problem required only the selection of division as the appropriate mathematical operation and a quick estimate of the quotient of approximately $2.00 divided by 20 for the price per ounce.

The item in Figure 4 was one of the most difficult on the NALS assessment. It asked participants to look at an advertisement for a home equity loan and calculate the total amount of interest charges associated with the loan. To correctly answer this item, participants had to multiply $156.77

Unit price
11.8¢ per oz.
You pay
1.89
rich chnky pnt bt
10693
16 oz.

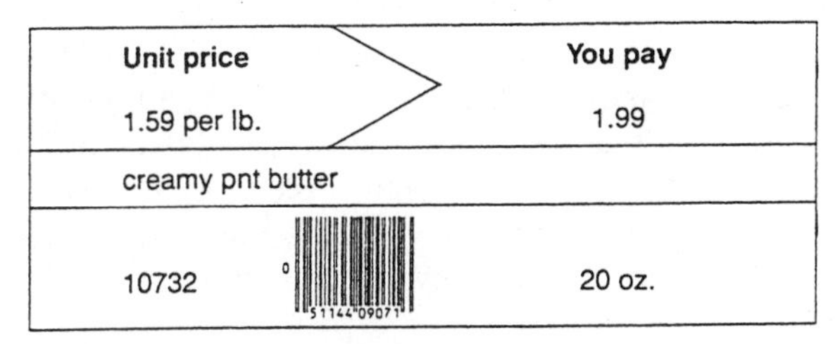

Figure 3.
Level IV quantitative literacy item from NALS.

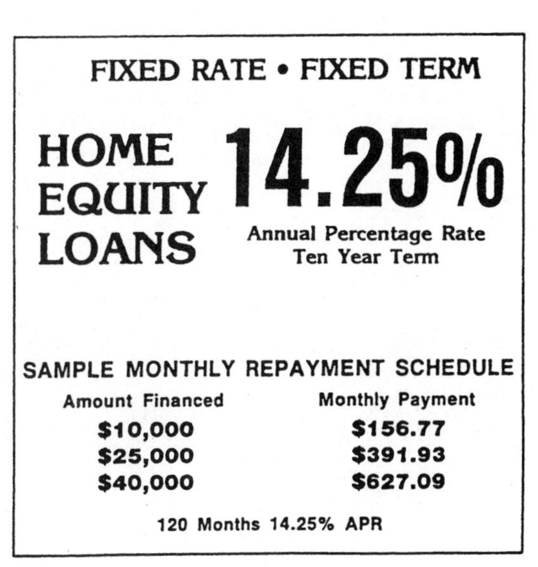

Figure 4.
Level V quantitative literacy item from NALS.

times 120 months to get the total costs and subtract the $10,000 loan value to obtain the total interest costs of $8,812.40. Only 4 percent of the nation's adults correctly answered items similar to this in content and difficulty.

The graphs in Figure 5 present the results of the NALS assessment of adult literacy. This carefully conducted random survey of more than 26,000 adults age 16 to 64 indicated that there is a considerable distance to go before all adults achieve full literacy as defined by this assessment.

Not surprisingly, analysis of the NALS findings showed a high correlation between the ability to read English and to achieve high scores on the document and quantitative literacy scales. However, the NALS test was administered individually to each person in the sample. Thus, while language is certainly an important factor in document and quantitative literacy, NALS buffers this to some extent by individual administration. Language plays an important (and ever-increasing) role in quantitative literacy as defined in this article. Nevertheless, there are important aspects of quantitative literacy that may be relatively language-free.

International Survey of Adult Literacy

At approximately the same time the NALS study was conducted in the United States, a companion study of eight nations was done under the auspices of the International Adult Literacy Survey (IALS), conducted by the Organization for Economic Cooperation and Development (OECD), the European Union, and UNESCO. This study, which included Canada, Germany,

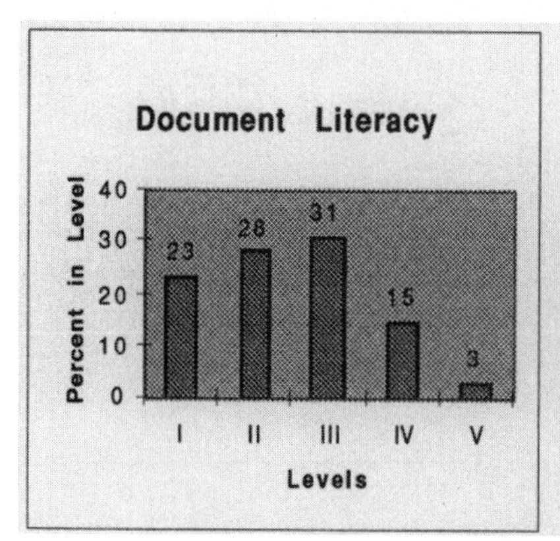

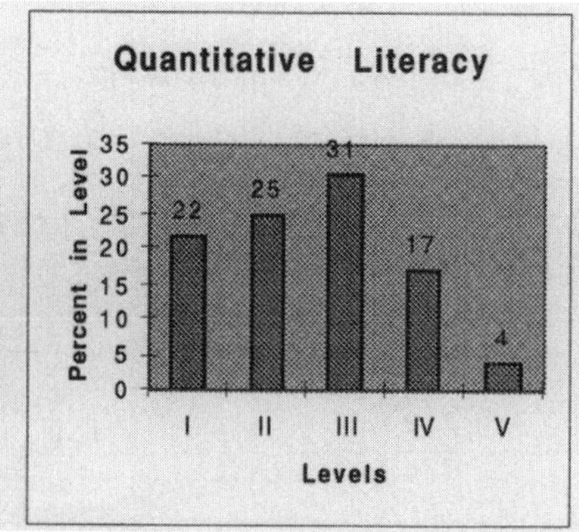

Figure 5.
Literacy levels and average literacy proficiencies from NALS.

Ireland, the Netherlands, Poland, Sweden, Switzerland, and the United States, used the same three literacy scales as the previous studies—prose, document, and quantitative. The latter two scales were subdivided into levels similar to those described in Table 1. The data in Figure 6 show the percentages of adults in each country performing at the top two levels of document and quantitative literacy.[16]

Analysis of these data by levels indicates that the percentages of adults functioning at the upper two document and quantitative literacy levels varies substantially from country to country. Sweden, with 36 percent on both scales, has a considerably higher percentage of its citizens functioning in the higher scale ranges than does Poland with only 6 percent and 7 percent. Education and environmental context variables, as well as other factors, easily explain these differences. Distributions by country show far fewer differences when the data are disaggregated by level of education and level of literacy expected for particular occupations.[17]

The IALS study differed from the NALS in that the assessment avoided multiple-choice questions and made heavier use of contextually embedded items. Four items, indicative of Level IV and V performance on the document and quantitative scales, are shown in Figures 7 to 10. The item in Figure 7, at Level IV on document literacy, asked respondents to summarize how the percent-

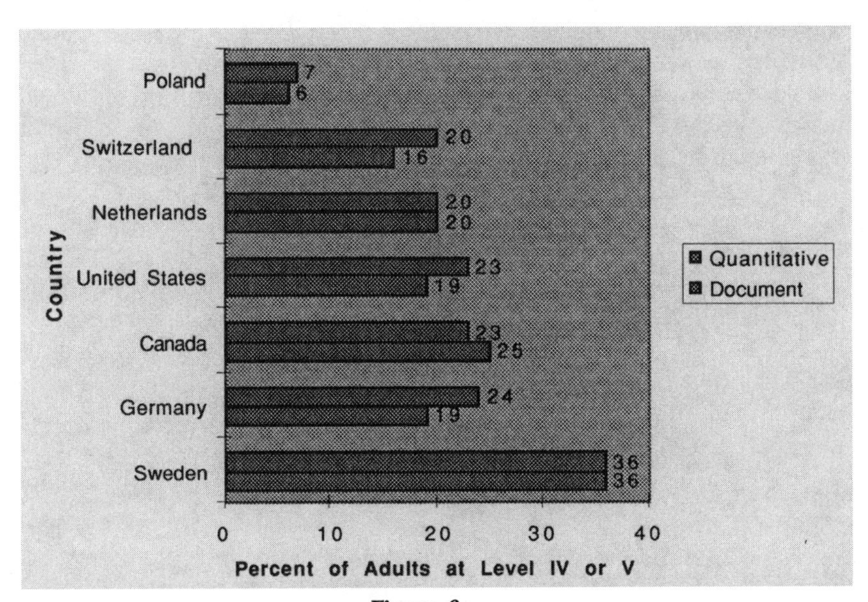

Figure 6.
Percentages of adult population by country functioning at Levels IV and V in document and quantitative literacy from IALS.

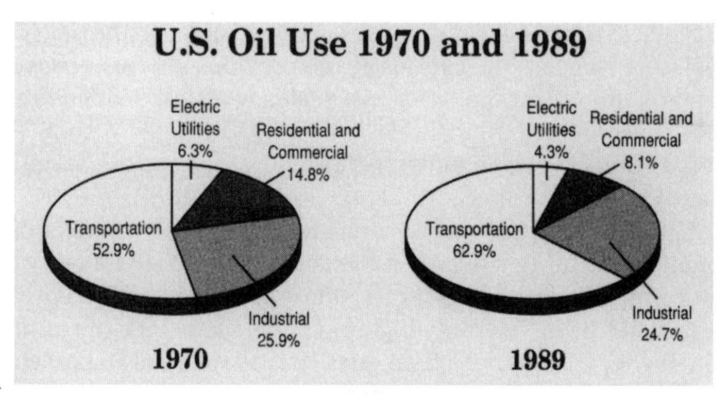

Figure 7.
Level IV document literacy item from IALS.

ages of oil used for different purposes changed over the period specified in the two pie charts. This required the respondent to study the two charts, comparing the percentages of oil used during the two years and then constructing a statement detailing those changes across the four fuel uses listed.

The item in Figure 8 measured performance at document literacy Level V. Here the respondent had to give two responses. The first was to designate the basic clock radio receiving the highest overall rating. The second was to indicate which full-featured radio was rated highest on performance. Both of these items reflect knowledge and skills that all citizens should have to make economically practical decisions affecting their lives and livelihoods. The data, however, indicate that fewer than one in five U.S. adults could function at either of these levels of document literacy.

The item in Figure 9, tied to Level IV on the quantitative literacy scale, asked respondents to calculate the number of kilometers traveled on a trip from Guadalajara to Tecomán and then to Zamora. The item in Figure 10, at Level V on the quantitative scale, was one of the most difficult items. Respondents were asked to determine the number of calories in a Big Mac that come from total fat. In the question, they were told that a gram of fat has 9 calories. Hence, they had to convert the number of grams of fat to calories and then calculate this number as a percentage of the total calories in a Big Mac. The results show that only one-third of U.S. adults are able to perform at the levels of quantitative literacy required by these two items.

Mathematical Literacy

Other analyses have been done of adult quantitative literacy. These assessments are probably better viewed as measuring mathematical literacy, since

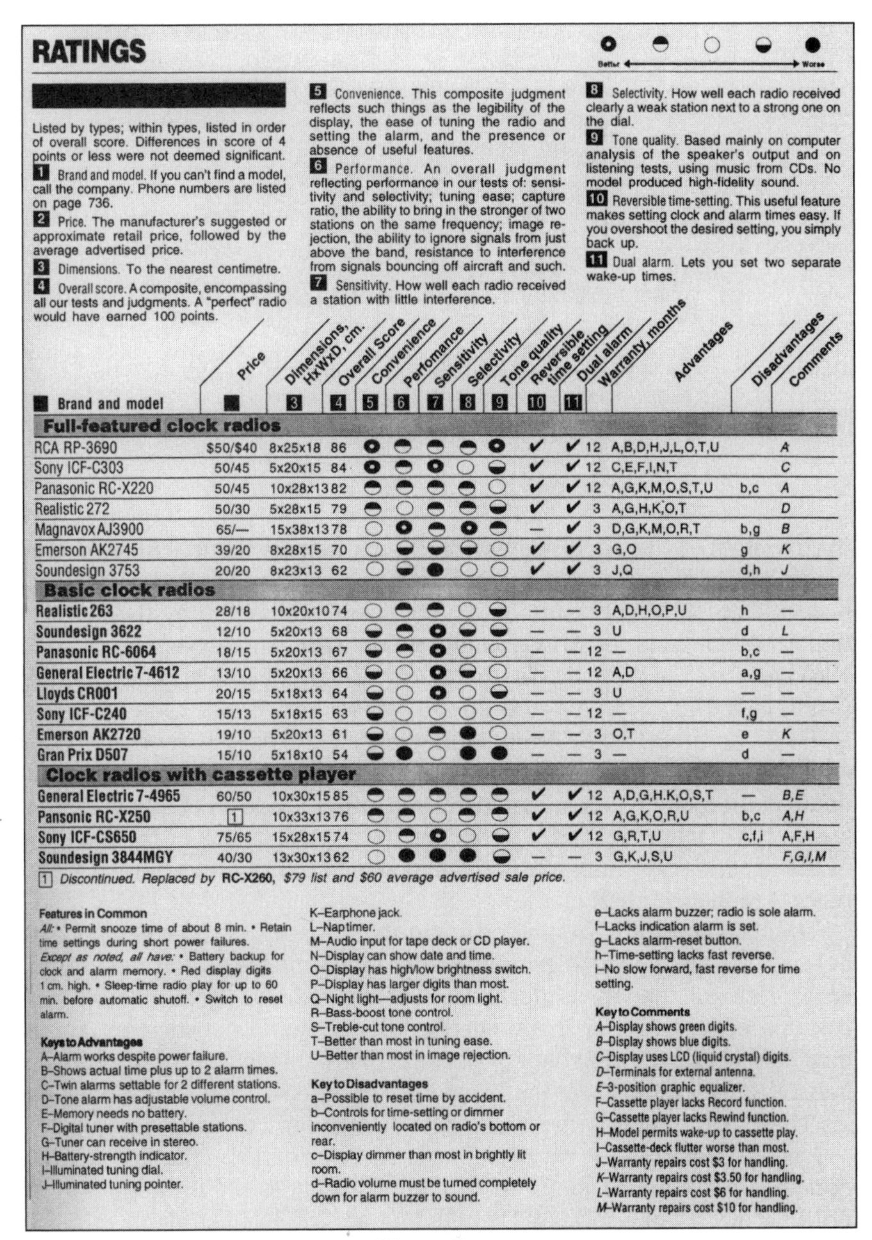

RATINGS

⊙ ◓ ○ ◒ ● (Better ← → Worse)

Listed by types; within types, listed in order of overall score. Differences in score of 4 points or less were not deemed significant.

1 Brand and model. If you can't find a model, call the company. Phone numbers are listed on page 736.

2 Price. The manufacturer's suggested or approximate retail price, followed by the average advertised price.

3 Dimensions. To the nearest centimetre.

4 Overall score. A composite, encompassing all our tests and judgments. A "perfect" radio would have earned 100 points.

5 Convenience. This composite judgment reflects such things as the legibility of the display, the ease of tuning the radio and setting the alarm, and the presence or absence of useful features.

6 Performance. An overall judgment reflecting performance in our tests of: sensitivity and selectivity; tuning ease; capture ratio, the ability to bring in the stronger of two stations on the same frequency; image rejection, the ability to ignore signals from just above the band, resistance to interference from signals bouncing off aircraft and such.

7 Sensitivity. How well each radio received a station with little interference.

8 Selectivity. How well each radio received clearly a weak station next to a strong one on the dial.

9 Tone quality. Based mainly on computer analysis of the speaker's output and on listening tests, using music from CDs. No model produced high-fidelity sound.

10 Reversible time-setting. This useful feature makes setting clock and alarm times easy. If you overshoot the desired setting, you simply back up.

11 Dual alarm. Lets you set two separate wake-up times.

1 Brand and model	2 Price	3 Dimensions, HxWxD, cm.	4 Overall Score	5 Convenience	6 Perfomance	7 Sensitivity	8 Selectivity	9 Tone quality	10 Reversible time setting	11 Dual alarm	Warranty, months	Advantages	Disadvantages	Comments
Full-featured clock radios														
RCA RP-3690	$50/$40	8x25x18	86	⊙	◓	◓	◓	⊙	✔	✔	12	A,B,D,H,J,L,O,T,U		*A*
Sony ICF-C303	50/45	5x20x15	84	⊙	◓	⊙	○	◒	✔	✔	12	C,E,F,I,N,T		*C*
Panasonic RC-X220	50/45	10x28x13	82	◓	◓	◓	◓	○	✔	✔	12	A,G,K,M,O,S,T,U	b,c	*A*
Realistic 272	50/30	5x28x15	79	◓	○	◓	◓	◒	✔	✔	3	A,G,H,K,O,T		*D*
Magnavox AJ3900	65/—	15x38x13	78	○	⊙	◓	⊙	◓	—	✔	3	D,G,K,M,O,R,T	b,g	*B*
Emerson AK2745	39/20	8x28x15	70	○	◒	◒	◒	○	✔	✔	3	G,O	g	*K*
Soundesign 3753	20/20	8x23x13	62	○	◒	●	○	○	✔	✔	3	J,Q	d,h	*J*
Basic clock radios														
Realistic 263	28/18	10x20x10	74	○	◓	◓	○	◒	—	—	3	A,D,H,O,P,U	h	—
Soundesign 3622	12/10	5x20x13	68	◒	◓	⊙	◒	◒	—	—	3	U	d	*L*
Panasonic RC-6064	18/15	5x20x13	67	◒	◓	⊙	○	○	—	—	12	—	b,c	—
General Electric 7-4612	13/10	5x20x13	66	◒	○	⊙	◒	○	—	—	12	A,D	a,g	—
Lloyds CR001	20/15	5x18x13	64	◒	○	⊙	○	◒	—	—	3	U	—	—
Sony ICF-C240	15/13	5x18x15	63	◒	○	○	○	○	—	—	12	—	f,g	—
Emerson AK2720	19/10	5x20x13	61	◒	○	◓	●	○	—	—	3	O,T	e	*K*
Gran Prix D507	15/10	5x18x10	54	◒	●	○	●	●	—	—	3	—	d	—
Clock radios with cassette player														
General Electric 7-4965	60/50	10x30x15	85	◓	◓	◓	◓	◓	✔	✔	12	A,D,G,H.K,O,S,T	—	*B,E*
Pansonic RC-X250	[1]	10x33x13	76	◓	◓	○	◓	◓	✔	✔	12	A,G,K,O,R,U	b,c	*A,H*
Sony ICF-CS650	75/65	15x28x15	74	○	◓	⊙	○	◒	✔	✔	12	G,R,T,U	c,f,i	A,F,H
Soundesign 3844MGY	40/30	13x30x13	62	○	●	●	●	◒	—	—	3	G,K,J,S,U		*F,G,I,M*

[1] *Discontinued. Replaced by* **RC-X260**, *$79 list and $60 average advertised sale price.*

Features in Common

All: • Permit snooze time of about 8 min. • Retain time settings during short power failures.

Except as noted, all have: • Battery backup for clock and alarm memory. • Red display digits 1 cm. high. • Sleep-time radio play for up to 60 min. before automatic shutoff. • Switch to reset alarm.

Keys to Advantages

A–Alarm works despite power failure.
B–Shows actual time plus up to 2 alarm times.
C–Twin alarms settable for 2 different stations.
D–Tone alarm has adjustable volume control.
E–Memory needs no battery.
F–Digital tuner with presettable stations.
G–Tuner can receive in stereo.
H–Battery-strength indicator.
I–Illuminated tuning dial.
J–Illuminated tuning pointer.
K–Earphone jack.
L–Nap timer.
M–Audio input for tape deck or CD player.
N–Display can show date and time.
O–Display has high/low brightness switch.
P–Display has larger digits than most.
Q–Night light—adjusts for room light.
R–Bass-boost tone control.
S–Treble-cut tone control.
T–Better than most in tuning ease.
U–Better than most in image rejection.

Key to Disadvantages

a–Possible to reset time by accident.
b–Controls for time-setting or dimmer inconveniently located on radio's bottom or rear.
c–Display dimmer than most in brightly lit room.
d–Radio volume must be turned completely down for alarm buzzer to sound.
e–Lacks alarm buzzer; radio is sole alarm.
f–Lacks indication alarm is set.
g–Lacks alarm-reset button.
h–Time-setting lacks fast reverse.
i–No slow forward, fast reverse for time setting.

Key to Comments

A–Display shows green digits.
B–Display shows blue digits.
C–Display uses LCD (liquid crystal) digits.
D–Terminals for external antenna.
E–3-position graphic equalizer.
F–Cassette player lacks Record function.
G–Cassette player lacks Rewind function.
H–Model permits wake-up to cassette play.
I–Cassette-deck flutter worse than most.
J–Warranty repairs cost $3 for handling.
K–Warranty repairs cost $3.50 for handling.
L–Warranty repairs cost $6 for handling.
M–Warranty repairs cost $10 for handling.

Figure 8.
Level V document literacy item from IALS.

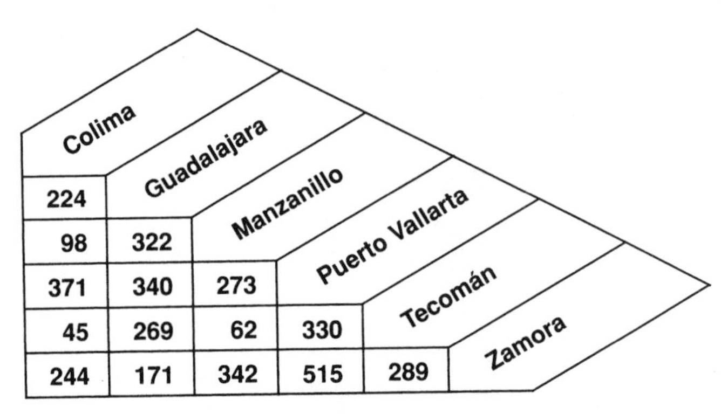

Figure 9.
Level IV quantitative literacy item from IALS.

they evaluated the effectiveness of the mathematics curriculum that is taught in school. In the United States, such studies were conducted by NAEP from 1973 to 1992; international studies were conducted by the International Association for the Evaluation of Educational Achievement (IEA) in 1963 and 1981[18] and the International Assessment of Educational Progress (IAEP) in 1991.[19]

These studies all assessed students' ability at the latter part of secondary education in mathematics. Across time, they consistently showed that students in the United States demonstrate a rather low level of mathematical literacy as defined in terms of performance on a test assessing the portions of the secondary school mathematics curriculum that teachers actually report teaching in their classes.

The 1963 and 1981 IEA studies found that those U.S. students who took at least four years of college preparatory mathematics performed considerably less well than their international counterparts on items dealing with calculus, college algebra, trigonometry, probability and statistics, and geometry. Moreover, IAEP eighth-grade international data indicated that the baseline for these low levels of performance in mathematics is in place by age 13. These findings offer disturbing signs for the future.

Pashley and Phillips[20] showed that these international findings are not equally apt comparisons for all students, and possibly for adults, in all portions of the United States. Their findings indicate that students in states with few urban centers (where social factors create deficient learning situations) may perform at levels comparable to those of students in countries ranked near the top in international comparisons of mathematical achievement.[21]

	Serving Size	Calories	Protein (g)	Carbohydrates (g)	Total Fat (g)	Saturated Fat (g)	Monounsaturated Fat (g)	Polyunsaturated Fat (g)	Cholesterol (mg)	Sodium (mg)
Sandwiches										
Hamburger	102 g	255	12	30	9	5	1	3	37	490
Cheeseburger	116 g	305	15	30	13	7	1	5	50	725
Quarter Pounder®	166 g	410	23	34	20	11	1	8	85	645
Quarter Pounder® w/Cheese	194 g	510	28	34	28	16	1	11	115	1110
McLean Deluxe™	206 g	320	22	35	10	5	1	4	60	670
McLean Deluxe™ w/Cheese	219 g	370	24	35	14	8	1	5	75	890
Big Mac®	215 g	500	25	42	26	16	1	9	100	890
Filet-O-Fish®	141 g	370	14	38	18	8	6	4	50	730
McChicken®	187 g	415	19	39	19	9	7	4	50	830
French Fries										
Small French Fries	68 g	220	3	26	12	8	1	2.5	0	110
Medium French Fries	97 g	320	4	36	17	12	1.5	3.5	0	150
Large French Fries	122 g	400	6	46	22	15	2	5	0	200
Salads										
Chef Salad	265 g	170	17	8	9	4	1	4	111	400
Garden Salad	189 g	50	4	6	2	1	0.4	0.6	65	70
Chunky Chicken Salad	255 g	150	25	7	4	2	1	1	78	230
Side Salad	106 g	30	2	4	1	0.5	0.2	0.3	33	35
Croutons	11 g	50	1	7	2	1.3	0.1	0.5	0	140
Bacon Bits	3 g	15	1	0	1	0.3	0.2	0.5	1	95

Soft Drinks

	Coca-Cola Classic®				diet Coke®				Sprite®			
	Small	Medium	Large	Jumbo	Small	Medium	Large	Jumbo	Small	Medium	Large	Jumbo
Calories	140	190	260	380	1	1	2	3	140	190	260	380
Carbohydrates (g)	38	50	70	101	0.3	0.4	0.5	0.6	36	48	66	96
Sodium (mg)	15	20	25	40	30	40	60	80	15	20	25	40

Figure 10.
Level V quantitative literacy problem from IALS.

This wide diversity in achievement on questions measuring algebra, geometry, and calculus offers mixed signals about student achievement in mathematics in school settings. Data from the NAEP trend studies, employing the same tests since 1973,[22] indicate a slight drop in the performance of 17-year-olds from 1973 to 1982, followed by a steady increase from 1982 to the present, with all of the decrease regained. At the same time, the NAEP national assessments in 1990 and 1992[23] contain several items on which students were unable to perform adequately. Many of these items required quantitative literacy performance reflecting consumer-related mathematics skills. For example, fewer than 1 in 20 twelfth-grade students could correctly answer an item asking them to apply the definition of a proposed income tax rate.

Increased rigor in the mathematics requirements enacted by states for public school students[24] provides hope for future improvement in school mathematics performance. Analyses of NAEP data seem to indicate that increased emphasis on mathematics results in higher levels of performance. However, there is a danger that such a performance gain could be based on successful solving of abstract items devoid of meaning in and transferability to real-world contexts. Care must be taken to ensure that efforts to improve students' quantitative literacy result in true gains in understanding—gains that reflect both a knowledge of and ability to apply mathematics in ways that empower understanding and broaden career options.

Endnotes

1. National Commission on Excellence in Education, *A Nation at Risk: The Imperative for Educational Reform* (Washington, D.C.: U.S. Government Printing Office, 1983).
2. Secretary's Commission on Achieving Necessary Skills, *Learning a Living: A Blueprint for High Performance* (Washington, D.C.: U.S. Department of Labor, 1992).
3. A. E. Lapointe, N. A. Mead, and J. M. Askew, *Learning Mathematics* (Princeton, N.J.: Educational Testing Service, 1992); C. McKnight, F. J. Crosswhite, J. A. Dossey, E. Kifer, J. O. Swafford, K. J. Travers, and T. J. Cooney, *The Underachieving Curriculum: Assessing U.S. School Mathematics from an International Perspective* (Champaign, Ill.: Stipes, 1987); I. V. S. Mullis, J. A. Dossey, J. R. Campbell, C. A. Gentile, and A. S. Latham, *National Assessment of Educational Progress 1992: Trends in Academic Progress* (Washington, D.C.: U.S. Department of Education, National Center for Education Statistics, 1994).
4. National Council of Teachers of Mathematics, *Curriculum and Evaluation Standards for School Mathematics* (Reston, Va.: National Council of Teachers of Mathematics, 1989).
5. L. A. Steen and S. L. Forman, "Mathematics for Work and Life," in *Prospects for School Mathematics,* ed. I. M. Carl (Reston, Va.: National Council of Teachers of Mathematics, 1995), 219–241.
6. *Learning a Living.*
7. I. S. Kirsch and A. Jungeblut, *Literacy: Profiles of America's Young Adults* (Princeton, N.J.: Educational Testing Service, 1986).
8. National Science Foundation, *Human Talent for Competitiveness* (Washington, D.C.: National Science Foundation, 1987).
9. *Literacy: Profiles of America's Young Adults.*
10. I. V. S. Mullis, J. A. Dossey, E. H. Owen, and G. W. Phillips, *National Assessment of Educational Progress 1992: Mathematics Report Card for the Nation and the States* (Washington, D.C.: U.S. Department of Education, National Center for Education Statistics, 1993); *NAEP 1992 Trends in Academic Progress.*
11. T. Carpenter, T. G. Coburn, R. E. Reys, and J. W. Wilson, *Results from the First Mathematics Assessment of the National Assessment of Educational Progress* (Reston, Va.: National Council of Teachers of Mathematics, 1978).
12. I. S. Kirsch, A. Jungeblut, L. Jenkins, and A. Kolstad, *Adult Literacy in America: A First Look at the Results of the National Adult Literacy Survey* (Washington, D.C.: U.S. Department of Education, National Center for Education Statistics 1993), 2.
13. Id., 3
14. Id., 11

15. Id., 92-93.
16. Organization for Economic Cooperation and Development, *Literacy, Economy, and Society* (Ottawa, Canada: Statistics Canada, 1995).
17. P. E. Barton, *Earning and Learning* (Princeton, N.J.: Educational Testing Service, 1989); P. E. Barton, *Becoming Literate About Literacy* (Princeton, N.J.: Educational Testing Service, 1994); and P. E. Barton and A. Lapointe, *Learning by Degrees: Indicators of Performance in Higher Education* (Princeton, N.J.: Educational Testing Service, 1995).
18. T. Husen, ed., *International Study of Achievement in Mathematics: A Comparison of Twelve Countries* (Stockholm, Sweden: Almqvist and Wiksell, 1967); *The Underachieving Curriculum.*
19. *Learning Mathematics.*
20. P. J. Pashley and G. W. Phillips, *Toward World-Class Standards: A Research Study Linking International and National Assessments* (Princeton, N.J.: Educational Testing Service, 1993).
21. National Science Foundation, *Indicators of Science and Mathematics Education* (Arlington, Va.: National Science Foundation, 1996); J. A. Dossey, P. B. Duckett, G. Lappan, R. J. Coley, R. M. Logan, and N. A. Mead, *How Do U.S. Students Measure Up?* (Princeton, N.J.: Educational Testing Service, 1994).
22. *NAEP 1992 Trends in Academic Progress.*
23. *Mathematics Report Card for the Nation and the States.*
24. R. K. Blank and D. Gruebel, *State Indicators of Science and Mathematics Education: 1995* (Washington, D.C.: Council of Chief State School Officers, 1995).

Thinking Quantitatively About Science

F. JAMES RUTHERFORD
American Association for the Advancement of Science

Science literacy depends on quantitative literacy, which in turn requires mathematics in context, often scientific or technological. In school, context is too often removed, leaving only the empty shell of mindless computation with "naked numbers."

Let me set the stage with an end-of-the-chapter math problem: A worm traveled 2 meters in 3 hours. What was its speed? From the science perspective, there is more to this (not altogether atypical) "real-world" problem than simply deciding what arithmetic operation to perform using the numbers 2 and 3 (and knowing how to carry it out correctly). A quantitatively literate person would respond not with an answer but with questions:

- What are the quantities we are dealing with? Was 2 meters the total distance traversed along the actual path of the worm, or was it the net displacement? Is 3 hours the total elapsed time, or the cumulative time when the worm was in motion? Without this information, *any* numerical answer is ambiguous at best.
- How precise is our knowledge of the quantities? It is hard to believe that for such a short distance we would measure only to the nearest meter (instead of to the nearest centimeter or millimeter) or to the nearest hour (instead of the nearest minute or second). Without knowing this, we cannot know how to express an answer. After all, 1 m/h, 0.7 m/h, 0.67 m/h, 0.667 m/h, etc., are different quantities, and hence different answers to the question.
- What is the point? The minimum possible speed with which that particular worm moved on one occasion, or what its top speed is, or how fast it moves under different circumstances, or how fast it generally moves? Only the first case—clearly the least interesting, if one is curi-

ous about worms—could possibly be answered with the data given. Or is the point how fast worms move—a very different matter requiring more information than is given? Without knowing what we are trying to find out, there is no way of knowing whether the information given is sufficient, or even whether the numerical problem is interesting enough to bother solving at all.

- What do we already know about the movement of worms? If the answer is nothing, then nearly any numerical answer will do (as long as the units properly express speed). Only if we know enough about worms (and how long a meter is!) to imagine what might be within the realm of possibility will we be able to decide whether our answer makes sense.

Quantitative Literacy

It might be argued that these questions about the worm problem are simply extraneous. After all, the purpose of asking students do this kind of problem is to improve their ability to calculate speed, given distance and time, not to teach them about measurement and significant figures, let alone to inform them about worms. But if the substance and expression of the answer do not matter, then the numbers and referents in such problems can be anything (e.g., what is the speed of a jet aircraft that travels 1.234 feet in 5 days?) and we are left with a context-free arithmetic exercise in which *quantity* is beside the point. The worm problem might just as well, better, perhaps, be given as: What is the speed when distance is 2 and time is 3, or, better yet, as what is the speed s when distance is d and time is t?

Indeed, students are taught the speed formula on the assumption that they will then be able to apply it correctly in any appropriate context. But how are people supposed to recognize an appropriate context when they see one? And how are they supposed to know which quantities are the correct ones to use in a given context? Are light-years units of distance or time? What about microns? Is it okay to express the speed of an elevator as floors per minute? Are shares of stock sold per hour on the New York Stock Exchange and gallons of water per minute flowing at some point in the Colorado River measures of speed, even though distance is not involved? For that matter, is a year always the same amount of time? And so on.

The truth is that people have difficulty using context-free knowledge and skills in everyday situations. It is instructive that the two rates most people can manipulate with some mathematical ease have to do with wages and automobile speed. The more abstract the mathematical construct, the less likely it is to be used in the practical affairs of everyday life. Knowing, for instance, how to solve $a = b/c$ for b is of little help in dealing with financial or motion problems, say, if we do not realize that the relationship applies to the situation at hand, or if we do not know how to apply it. How do I know I am not dealing with $a = bc$? What is a and what is b and what is c? Is there or is there not a d? An e? What units have to be used?

If students learn to use specific rate relationships properly *in many different subjects,* there is a chance that one day they will be able to understand the generality of the idea of rate. Fortunately, the opportunities to do so are abundant: rate relationships are the bread and butter of the physical, biological, and social sciences, and they are not strangers in history, geography, business, music, physical education, and many other school subjects. Students who have learned to use the time-rate relationship in many different and meaningful contexts will be well on the road to quantitative literacy even if they never quite grasp the concept of rate in the abstract (which is to say, in all its mathematical grandeur). The converse is less likely.

Among the most significant of the contexts for developing quantitative literacy are those associated with science and technology. In our private lives and as concerned citizens, we need to make sense out of claims made, information disseminated, and solutions proposed by government agencies, private organizations, the media, political candidates and organizations, experts (acknowledged and self-proclaimed), interest groups, and friends. About what? The environment, natural resources, immigration, economic trends, health care, crime, consumer safety, investment in research, and more. To deal with such matters well, citizens need to possess certain basic mathematical capabilities *understood in association with* relevant scientific and technological knowledge.

As inquisitive human beings, moreover, we live in a marvelous time in which science is making great strides. Through science we are discovering what the universe is like and how it got that way, how our planet is constructed and how it behaves, how living organisms are alike and different, how they came into being and relate to one another, how the earth's species have changed over time, what humans are like as organisms and as a species able to create social arrangements and artifacts. However, to follow this grand adventure story—and what a shame for anyone to miss out on it—with enough understanding to share in the excitement of discovery, people must be quantitatively literate. Problem solving is not the only "practical" use of mathematics.

This argument that quantitative literacy is contextual has significant educational ramifications. It suggests, for one thing, that the starting place for deciding what constitutes quantitative literacy is less mathematics itself than the contexts in which people are most likely to encounter the need for mathematical insights and skills. It also leads to the radical possibility that much of the teaching of mathematics—something different from the application of mathematics already learned—ought to take place in subjects other than mathematics. The latter proposition is, however, not likely to be warmly embraced either by mathematics teachers (who take the teaching of mathematics to be their rather exclusive responsibility) or by the teachers of other subjects (who believe they have more than enough to teach already and lack training in teaching mathematics).

Science Literacy

Science literacy and quantitative literacy both have to do with preparation for ordinary life as adults, not with preparation for technical vocations. Hence, it is important to distinguish quantitative literacy in science from mathematical competency in science, the latter referring to the kinds and levels of mathematical knowledge and skill needed to practice science professionally (knowledge and skill requirements that vary from science to science and change continually).

But just what is science literacy? Recently, a broad-based consensus on what constitutes science literacy among adults has emerged in the scientific community. It was described in *Science for All Americans,*[1] a report produced under the auspices of the American Association for the Advancement of Science (AAAS). The view of science literacy outlined in that report was then reflected in *Benchmarks for Science Literacy,*[2] also done under the aegis of the AAAS, and in the *National Science Education Standards,*[3] developed under the guidance of the National Research Council of the National Academy of Sciences. The development of this consensus over a 10-year span engaged hundreds of scientists, mathematicians, and engineers, and thousands of K–12 teachers of science and mathematics. It is strongly supported by scientific societies and science teaching associations.

It seems justified, therefore, to take *Science for All Americans (SFAA)* to represent what, from the perspective of the scientific community, constitutes science literacy. In that view, scientifically literate people are:

- Well-informed about the nature of science, mathematics, and technology as related endeavors, including their similarities, connections, and differences, and aware of the historical impact of the scientific enterprise;
- Familiar with the views of the world—physical, biological, and social—depicted by current science, mathematics, and technology;

- Capable of using powerful thematic concepts that cut across all of science, mathematics, and technology, particularly those having to do with systems, models, constancy and change, and scale; and
- Possessed of habits of mind that are characteristic of science.

In terms of quantitative literacy, *SFAA* expects that scientifically literate persons will:

- Understand the nature of mathematics, its role in scientific inquiry and technological progress, and its dependence, in turn, on science and technology;
- Grasp sufficient mathematics to understand (at a level suitable for the educated public) important scientific and engineering concepts (such as those spelled out in *SFAA*) and certain themes (such as change and scale) that are not exclusively scientific but are nevertheless widely used in science and technology; and
- Possess quantitative skills sufficient for following scientific stories that appear in the media, participating in public dialogue on issues involving science and technology, and responding critically to claims made in the name of science and technology.

It is important to be clear on the limits of what is being claimed. These general properties of quantitative literacy emerge from our focus on the goal of science literacy. It does not necessarily follow that this is all there is to quantitative literacy. *SFAA* and *Benchmarks* develop the concept of quantitative literacy from the perspective of science in terms of what students should be expected to learn. The commentary that follows is meant to illustrate this scientific perspective on quantitative literacy.

The Nature of Mathematics

Knowing about mathematics is a very different matter from knowing mathematics. It is not surprising that scientists would like people to be good enough in mathematics to understand scientific concepts and processes, but why would they care whether or not people know about mathematics as such, i.e., about what mathematics *is* and how it works? The answer is in part self-interest—understanding the nature of mathematics is a necessary condition for understanding the nature of science. And it is in part a matter of definition, since, according to *SFAA*, science literacy calls for an understanding of the *scientific endeavor as a whole*, not simply of science in isolation.

To begin with, it is important to overcome the widespread belief—fostered all too often by the way mathematics is taught in schools (to all students) and in colleges (to nonmajors)—that mathematics is only another name for computation, albeit computation that gets fancier and harder and (truth be known) less useful as we move up the chain from sums to long division to trigonometry to calculus and beyond. This misconception interferes with seeing mathematics as a science in its own right, as a science of abstract patterns and relationships that may sometimes be quantitative but can also be logical, spatial, temporal, functional, or, as is often the case, a combination of such forms.

As a theoretical discipline, mathematics is in principle unconcerned with whether the abstractions it explores have counterparts in the real world, whereas applied mathematics focuses on solving practical, real-world problems. Yet the same general approach is used in both theoretical and applied mathematics: represent some aspects of things abstractly, manipulate the abstractions by rules of logic to find new relationships among them, and see whether the new relationships say something interesting about the original things, whether the original things are themselves abstractions or actual objects or events. History makes clear a stunning and unpredictable result: time and time again, the discoveries of theoretical mathematics turn out to have practical value, and those of applied mathematics to have theoretical value.

This Janus-like character of mathematics helps explain the long and mutually beneficial alliance between science and mathematics. Science provides mathematics with interesting real-world problems to investigate, and mathematics provides science with powerful analytical tools and with a universal language for expressing its findings. In short, science and mathematics are both trying to discover general patterns and relationships, and in doing so they lean on each other. The implication of this for quantitative literacy is that students need to learn how mathematics goes about its business, why theoretical mathematics is at least as important as applied mathematics, and what the connections are between science and mathematics.

Computation and Estimation

This, of course, is not to say that computation is insignificant as a component of quantitative literacy. Quite the contrary. While computation in one form or another is rarely absent in the doing of science (and, as such, mainly of professional interest), it is also usually embedded in descriptions of scientific concepts, discoveries, and claims (and, as such, is necessary for the public understanding of science).

The computational skills needed for science literacy are little different from those needed for everyday life, and they are not overwhelming. At least, that is the conclusion reached in *Science for All Americans.* Everyone needs to memorize and be able to recall immediately such number facts as the sums, differences, and products of whole numbers from 1 to 10, the decimal equivalents of common fractions, the relation between fractions and percentages, and the commonly encountered powers of 10, and to be able to add any two 2-digit numbers mentally and to multiply and divide mentally any number by 2, 10, and 100 to one or two significant figures. Nothing new there, except to note that while it calls for less than the schools try to teach, it is more than many adults are actually able to do.

In any case, such basic number skills, while necessary, do not add up to quantitative literacy. Calculator skills are also needed to take people past mental arithmetic to real numeracy. For all the faith people have in paper-and-pencil calculations, the fact is that such calculations are, as claimed in *SFAA,* "slow, prone to error, and as conceptually mysterious to most users as are electronic tools. When precision is desired, when numbers being dealt with have multiple digits, or when the computation has several steps, the calculator offers many practical advantages."[4] The particular calculator skills called for in *SFAA* include:

- Adding, subtracting, multiplying, and dividing any two whole or decimal numbers (but not powers, roots, or trigonometric functions);
- Finding the decimal equivalent of any fraction;
- Calculating percentages and taking percentages of quantities;
- Finding reciprocals;
- Determining rates from magnitudes and magnitudes from rates;
- Calculating circumferences and areas of rectangles, triangles, and circles, and volumes of rectangular solids;
- Finding the mean of a set of data;
- Determining the value of simple algebraic expressions by numerical substitution; and
- Converting compound units.

But calculators cannot make up for reasoning errors, so they often give a wrong answer because information was entered incorrectly or the wrong sequence of operations was used. Thus, quantitative literacy also requires estimation skills that enable people to check an answer for plausibility before accepting it. Such skills are also invaluable for assessing the probability of quantitative claims made by others (especially assertions appearing in both print and electronic media). Moreover, there are many situations in everyday life in which making an estimation is all that is needed. Doing precise computations is a waste of time. Making sound estimates does not come easily, however; this skill must be learned and practiced frequently in many different contexts.

Numbers and Quantities

From the standpoint of science, almost everything of interest is a quantity, a number connected to a label: 55 Mbytes, 6000 light-years, 141°K, 2.5 radians/sec. Science asks how big, how heavy, how far, how fast, how often, how hot, etc.—questions that cannot be answered with naked numbers. Ratios often appear as numbers sans units, but they have implied content (e.g., 3/4 may stand for 3 persons/4 persons). Even the simple numeric question, "how many," implies an answer such as 12 elk or 435 samples, not just 12 or 435.

Unfortunately, mathematics instruction all too often settles for drilling students in the use of numbers as abstract entities. It seems to be taken for granted that students will later know the difference between numbers and quantities and attach labels as needed to make sense out of something in the natural world. Science teachers at every level will testify that in fact students have great difficulty using whatever number skills they have to deal with actual quantities, or at least with the kinds of quantities encountered in the sciences. And the same can surely be said with regard to technological contexts.

Let us return to our worm, the one that, according to the end-of-chapter problem, raced 2 meters in 3 hours. The quantitatively literate person knows that the answer is not "6," because it has to be 6 of something. But 6 meter-hours does not seem to make sense, since speed has to be expressed in some distance/time unit. But 6 m/h is also out of the question because the creature could not have gone farther in 1 hour than it did in 3 hours, unless of course it took an indirect path, which raises a question about the validity of the quantities themselves—just how far did the worm actually travel? And for that matter, was it actually moving all the time? The quantitatively literate person knows that the way to get a speed unit is to divide distance by time, but will not accept 0.66666667 m/h as an answer to this problem, no matter what a calculator says 2/3 equals. The teacher should (1) accept only 0.7 m/h as correct, and (2) vow never again to assign that problem, or others like it, except, perhaps, to introduce ideas about number and quantities and their use.

Measurement

In science, quantification is not everything, but there is no denying its prominence. Hence measurement is never far away in science (and engineering), and thus, in the scientific view of the world, understanding the link between measurement and quantity is essential for quantitative literacy.

Conceptually, measurement is simple enough: it consists of counting how many there are of something or other. Count how many centimeters placed end to end it takes to match the length of an object. Count how many seconds, one after another, it takes for a particular event to happen. Count how many students take Algebra II each year. But of course in actual practice counting in the 1, 2, 3 . . . sense rarely occurs, for we use instruments to help *estimate* a count—meter sticks, clocks, survey forms. And there is the hitch, or double hitch: instruments and people are both fallible counters, and most things, unlike, say, a small bag of marbles, do not present themselves as a collection of separately discernible whole entities. (In fact, different counts of a large bag of marbles are very likely to give different results.)

Thus, from the perspective of science, quantitative literacy includes understanding the nature, importance, and limitations of measurement; expecting quantities to be expressed in terms of numbers and labels that reveal precisely how and in what units they were measured (2.00 m, not 2 m, if the displacement of the worm was measured to the nearest centimeter); and knowing how the expression of calculated quantities relates to the measured quantities on which they are based (worm speed of 0.67 m/h, given distance measured in centimeters and time in minutes). To achieve such insights, students need to have extensive experience taking measurements in many different contexts—measurements that are used for computation that addresses some sensible question that students care about or that are used to critique the mathematical treatment of quantities by others. Such experience should be provided in all classes where quantitative data are used—in science and social studies, of course, but also (and especially) in mathematics, the natural refuge of measurement-free numbers.

In everyday life, as well as in science, quantitative literacy must surely imply having some facility with the common instruments that produce usable quantities. Everyone ought to be able to use appropriate instruments to make and properly label direct measurements of length, volume, weight, time intervals, and temperature, and to take readings from standard meter displays, both analog and digital. The "hands-on" movement in mathematics, led by the National Council of Teachers of Mathematics, now legitimizes the introduction of measurement at all levels of mathematics instruction.

Relationships

Science is more often interested in the relationship between quantities or categories of quantities than in single quantities or categories—between distance and time, economic status and test scores, altitude and temperature, prey and predators, for instance. Fortunately, mathematics makes available useful ways to express quantitative relationships, a basic set of which skills should be in everyone's intellectual toolbox. Mathematics also provides ways to express some important geometric and logical relationships that need to be part of that toolbox. In particular, everyone should be able to:

- Organize information into simple tables and read such tables with understanding;
- Read simple graphs (circle, bar, and line), and depict relationships by drawing freehand graphs to show trends (steady, accelerated, diminishing-return, cyclic, stepwise);
- Check the correspondence between tabular, graphic, and verbal descriptions of quantitative relationships;
- Use symbols to represent quantities and manipulate simple algebraic statements;
- Use basic geometric ideas, including perpendicular, parallel, similar, congruent, tangent, rotation, and symmetry, to describe spatial relationships; and
- Distinguish between the logical relationships expressed in the terms "if . . . then . . .," "and," "every," "not," "correlates with," and "causes." (Needless to say, the widespread confounding of the last two terms is especially distressing to the scientific community.)

These skills also require some understandings. For example, an equation can be a statement either of an actual quantitative relationship or of a logically possible relationship. The generality of the former can be established only by finding more instances of it, whereas the latter can be proved in the mathematical sense, but whether it holds in the real world of experience can be determined only by subjecting it to empirical verification. A related understanding helps explain the application of mathematical relationships to the real world. Even though real objects are never perfectly triangular, rectangular, round, etc., known mathematical properties of such figures can be usefully applied to them, provided their approximation to the ideal geometric figures is close enough.

Finally, everyone should understand that large changes in scale often change the relationship between quantities. Perhaps the most common instance of this in science, engineering, and everyday life is the relationship of volume to surface area: as objects increase in size, their volume increases faster than their surface area. Thus properties such as weight that depend on volume change disproportionately in comparison with properties such as strength that depend on surface area.

Uncertainty

Science and technology rarely deal with certainty. To be sure, science is engaged in trying to find out how the world works, and to the degree that it succeeds, so does our ability to describe with confidence what has happened in the past and to predict accurately what is likely to happen in the future. But scientific knowledge is rarely, if ever, absolute, not to mention sometimes being wrong, and so there always remains some uncertainty in even the most scientific of explanations and predictions. Engineering has both scientific knowledge and craft experience to draw on, but hardly fares better—bridges still collapse unexpectedly—if for no other reason than that engineering projects typically encompass more variables than can be dealt with simultaneously, or because decisions have to be made even in the absence of sufficient information.

This is not to say that science is anybody's guess and that engineering is irresponsible, but that unlike theoretical mathematics, science and engineering are by nature probabilistic. In fact, good science now requires scientists to make probability estimates, whenever possible, of the validity of their data and conclusions and to report them along with their findings, and good engineering requires engineers to estimate and report the risks (which are probability statements) associated with particular designs. Not surprisingly, therefore, statistics and probability are highly valued in all the sciences and in engineering—and not only to prepare science and engineering majors for their life's work. From the perspective of science literacy, some knowledge of statistics and probability is essential for everyone.

With regard to statistics, nothing may be more important than understanding the nature of aggregated information. "Average" may be the world's most frequently abused term, and is, in any case, no stranger in discussions of those public issues—resource distribution, economic trends, health care delivery, the environment, species preservation, research funding, and the like—to which knowledge from the natural and social sciences is relevant and that invariably invoke the use of claims based on the analysis of large amounts of data.

When information is summarized by plotting the data along a number line, it becomes relatively easy to see what is happening: where the data pile up, where some data are separate from others, what the extreme values are, and how much information there actually is on which to base a judgment. But in most contexts such detail is inconvenient. Unfortunately, when data are summarized numerically, most of the story gets lost. Quantitatively literate individuals know enough statistics to raise the right questions when confronted with assertions about this or that aspect of the world based on averages. Are the data on which the purported average is based distributed normally or skewed to one side or the other? How much variation is there around the central tendency? Was that central tendency calculated by dividing all values by the number of entries? Or is it the most common value of all the entries? Or is it the value that divides the data into two equal collections of entries? On how much data is the average (mean, mode, or median) based? Where did the data come from? And so forth.

As to probability, it is important for everyone to understand that it is a mathematical way of coping thoughtfully with the uncertainty that is a common feature of the real world, the world of interest to science and technology. Some uncertainty is built into nature, as expressed so eloquently in Heisenberg's Uncertainty Principle. Some is due to the physical impossibility of gathering all the information we might want, for example, the velocity of each and every molecule in a sample of something, or the opinion of every person in the country on some issue. Uncertainty may also result from lack of precision in data collection or from the lack of models by which to combine the information meaningfully.

Probability is useful because it puts boundaries on uncertainty and moderates claims, suggesting how confident we should be about the claims being made. It moves people away from a demand for absolute knowledge, something that science is not in the business of delivering, or for risk-free technologies, something engineers cannot guarantee. From the perspective of science, therefore, quantitative literacy must include the idea of uncertainty itself as it is manifested in various scientific and engineering contexts. Science teachers (especially in high school and college courses) should not shy away from discussing the limits of knowledge (as well as, of course, the growth of knowledge), and, where possible, showing how those limits can be expressed mathematically.

People usually encounter probability in the form of predictions—the probability that some event (or sequence of events) will happen. These probabilities are estimated using two very different methods: one based on past experience and accepted models of how things work (weather and economic forecasts are typical), the other (as in lotteries, and the kind commonly taught in mathematics courses) based on the number of alternative outcomes that are possible, assuming that all possible outcomes are accounted for and are equally likely to happen. Quantitatively literate people will understand the difference between these two different methods of estimating probabilities.

Moreover, quantitatively literate people will be comfortable with the various numerical ways in which probabilities are expressed—as a point on a scale running from 0 to 1, as a percentage, as a fraction with any denominator, or as odds—and will be able to shift effortlessly from one form to another. Of course the trouble with all such numerical expressions of probability is that they can imply greater precision than they deserve, which is why the manner in which they were calculated needs to be explained. People should expect, for example, that empirical probability statements will be accompanied by information on the size of the sample on which they were based and how that sample was acquired.

Finally, there are common misinterpretations of expressed probabilities that are especially worrisome in science and technology. One is to infer the fate of a particular individual (organism, object, event) from a probability statement, rather than taking such a statement to apply only to the fate of a proportion of a population of all such individuals. Another is to fail to distinguish between proportion and actual count. The example given in *SFAA* will serve here: a medical test with a probability of being correct 99 percent of the time may seem highly accurate, but if that test were performed on a million people, approximately 10,000 individuals would receive false results (and of course it would not be known whether a particular person received correct or false results).

Critical-Response Skills

From every direction come claims that purport to be true. They appear in newspapers and magazines, on television and radio, and in the speeches and conversation of almost everyone. Often these claims, whether correct or not, take on the mantle of science, even if only by implication: "Most doctors say that . . . ," "It's scientific fact that . . . ," "Experiments show that . . ." "It has been proved that . . . ."

Science literacy does not require that people become knowledgeable enough in science and engineering to be able to critique the technical publications of experts in those fields; it does call for them to be able, as stated in *SFAA*, "to detect the symptoms of doubtful assertions and arguments."[5]

Sometimes weak arguments reveal themselves by using celebrity as authority (not only movie stars and sports heroes, but on occasion no less a "scientific authority" than the United States Congress), intermingling facts and opinions in a way to suggest that the latter are the former, presenting explanations as the only possible ones worth consideration, or making no mention of control groups when basing a claim on "experimental evidence." Perhaps the most breathtaking of such arguments are those that dismiss contrary scientific views as "only theoretical" and hence not to be taken seriously, or those that imply that all members of a target group—say "teenagers," "immigrants," "scientists," or "Asians"—have certain characteristics that are not shared by members of other groups.

But at least as often, weak arguments give themselves away by using faulty mathematics. With regard to claims made in the media and elsewhere about how things work in the natural, designed, and social worlds, the problem may be less that they include calculation errors or use wrong algorithms than that they misuse logic, statistics, and graphing. Scientifically literate people, being quantitatively literate, are therefore on the lookout for such misleading practices as the following (taken from *SFAA*):

- Failing to make explicit the premises of an argument;
- Taking the converse of true statements to be automatically true (claiming that all eight-legged creatures are spiders because all spiders have eight legs);
- Asserting that one thing "causes" another just because it happens at about the same time (saying that the stock market rose in 1995 "because" unemployment went down that year);
- Reporting average results, but not the amount of variation around the average ("the average temperature on the planet Mercury is about 15°F," although it swings from about 300°F above to nearly 300°F below zero);
- Giving no information on the size of the sample from which an average was calculated (the infamous but still widely used ploy, "In a recent study, 90 percent of gastroenterologists recommended . . .");
- Mixing absolute and proportional quantities ("there were 3,400 more robberies in our city last year than the year before, whereas other cities had an increase of less than 1 percent");
- Displaying graphs that distort the appearance of results (by chopping off part of the scale, using unusual units, or using no scale units at all); and
- Extrapolating curves on a graph far beyond the data, or based on only a very few data points, or presented without a hint of how much data actually lie behind the graph.

Context Matters

We might plausibly argue that this science-based characterization of quantitative literacy is in no significant way idiosyncratic to science and technology. Whether it is or is not matters little, however, for the more important point is that quantitative literacy is an essential ingredient of science literacy, no matter what may be true in other domains. Stripped to the bones, our argument is simple:

- Science literacy is enormously important to the future of civilization;
- Science literacy requires that all people acquire certain knowledge and skills in science, mathematics, and technology; and
- The mathematics component of science literacy must be strongly contextual.

It is with regard to such a contextual relationship that the illustrations of quantitative literacy given here were selected. It seems unlikely that such literacy will result from the way in which science and mathematics are taught in schools and colleges today.

And now, with all that behind us, it is time for the next end-of-the-chapter problem. (Being an even-numbered problem, the answer will be given.) During a seven-year period, math scores for a certain group increased three points. How fast did the scores increase? (*Answer:* 0.7142857) For extra credit: At that rate, how long will it take students to become quantitatively literate? (*Answer:* About pi-square times as long as it would take a worm to circle the globe, or 132, whichever comes last.)

I wish to express my appreciation for the insightful advice provided me by Andrew Ahlgren and Gerald Kulm in writing this paper.

Endnotes

1. *Science for All Americans* (Washington, D.C.: American Association for the Advancement of Science, 1989).
2. *Benchmarks for Science Literacy* (Washington, D.C.: American Association for the Advancement of Science, 1995).
3. *National Science Education Standards* (Washington, D.C.: National Research Council, 1996).
4. *Science for All Americans.*
5. Ibid.

Mere Literacy Is Not Enough

GEORGE W. COBB
Mount Holyoke College

As computers enhance the value of verbal, visual, and logical skills needed to reason with data, more than ever citizens and employees need skills in quantitative reasoning that create a tapestry of meaning blending context with structure.

If you look on page 162 of the World Bank's 1995 *World Development Report,* you will find adult literacy rates for 132 countries. You will not find a similar list, however, for quantitative literacy. Quantitative literacy is much narrower in scope, but much harder to define, much harder to measure, and much harder to achieve. In this essay, I will examine the reasons for this anomaly, offer a framework for thinking about quantitative literacy, and express some concerns about the future.

But first, a quick look at the asymmetry. The *World Development Report*[1] has published adult literacy rates for 20 years. The U.N.'s *Human Development Report*[2] is a similar though newer series. It too publishes adult literacy rates, but nothing on quantitative literacy. The Organization for Economic Cooperation and Development (OECD's) 1995 *Literacy, Economy, and Society,*[3] a first effort of its kind, does report on quantitative literacy, but only for seven countries, all of them among the most developed in the world.

Why is it that for decades literacy rates have been recognized as important indicators of work force preparedness and socioeconomic strength, measured and reported for countries at all levels of economic development, yet comparable rates for quantitative literacy have only recently been measured in a way that permits cross-country comparisons, and even now only for a handful of the most developed countries?

For a short, simple answer, we need only one word: computers. However, this surface explanation conceals a second asymmetry with unsettling implications: quantitative thinking is becoming more important mainly because of computers, but these same computers are cheapening the value of many traditional quantitative skills—precisely the ones that are easiest to learn and least concentrated among an educated elite.

To clear the way for a closer look at these implications, we must first get out from under the phrase "quantitative literacy." This phrase suggests a parallel with ordinary literacy that erects a real barrier to understanding. Literacy, as defined and measured by various government agencies, reduces the answers to two simple questions: Can you read? Can you write? The phrase "quantitative *literacy*" tempts us to think of the analog for numbers: Can you count? Can you calculate? But these questions focus on the low end of a continuum inhabited by those skills whose value is quickly eroding. Value is increasing only at the upper end, and there *reasoning* is a better description than literacy.

Quantitative *reasoning* requires a difficult integration of four very different kinds of thinking. This makes it a kind of cognitive emulsion, an unstable Hollandaise of the intellect that constantly threatens to separate into its more basic forms of thought. One minor consequence of this uneasy structure is that any attempt to understand and define quantitative reasoning solely in terms of measurable skills is apt to overlook the dynamic tensions that hold the skills together. A far greater consequence is that essential elements of quantitative reasoning tend to be overlooked, misunderstood, and underemphasized in our school curricula. This underemphasis has costly consequences.

Numbers in Context

Having ducked literacy in favor of reasoning, I now turn to "quantitative" and its relation to mathematics. To some, mathematics is about numbers, and so to them quantitative means "mathematical." To others, mathematics is not so much about numbers as about patterns and their logical relationships. These patterns and their logic inhabit the bright world of Plato's ideal forms; numbers are merely the shadows cast by these forms on the wall of Plato's cave. All the same, the link to mathematics remains, and with it the temptation to see in quantitative some (possibly inferior) form of mathematical.

To equate quantitative reasoning with mathematics, however, is to succumb to blind-men-and-elephant disease. Mathematical thinking is essential; indeed, quantitative reasoning is supported by it, but to embrace mathematics as the essence of quantitative reasoning comes no closer to reality than did the blind man who hugged the elephant's leg. Something else besides mathematics, something quite different from it, is needed to link mathematical thinking to civic discourse. Plato knew this, of course. His curriculum began the education of future philosopher kings with a decade of mathematics, as a foundation for reasoning, but followed that decade with another devoted to other subjects—more difficult in Plato's estimation—that brought pure reason into contact with the real world.

How can we characterize the missing ingredient? Two millennia after Plato, the mathematician and religious apologist Blaise Pascal asserted that the entire spectrum of the intellectual enterprise could be reduced to just two pure forms of thinking, which he called *esprit de géométrie* and *esprit de finesse*. As described by Jacques Barzun in *The House of Intellect:*

> There are, said Pascal, two ways of thought, usually found in different minds. The geometric, or scientific mind works according to principles that are plain and few, but remote from common life. It may be hard to find these principles and grasp them from the very first, but once they are grasped one would have to be mentally deficient to confuse or misapply them. On the contrary, the subtle mind (*esprit de finesse*) works with principles that reside in the midst of life for everybody to see. But one needs a good eye to detect them because they are so numerous and so fine. It is almost impossible that some of them should not escape notice. Now, oversight inevitably leads to error, and hence one must be absolutely clear-eyed, and be a strict reasoner besides, if one does not want to go astray on well-known premises.[4]

To bring into sharpest focus just how it is that these two forms of thought come together in quantitative reasoning, let us elaborate on statistician David Moore's observation that "data are more than numbers; they are numbers with a context."[5] The contrasting, indeed opposing, roles of context in the mathematical and data analytic aspects of quantitative reasoning determine its most fundamental tension.

Although mathematics often relies on applied context for motivation and as a source of problems, ultimately, the focus of mathematics is on abstract patterns. Context is part of the irrelevant detail that must be boiled away over the flame of abstraction in order to reveal the previously hidden crystal of pure structure. *In mathematics, context obscures structure.* Like mathematicians, those who reason with data also rely on patterns, but ultimately, in quantitative reasoning, whether patterns have meaning, and whether they have any value, depends on context. *In quantitative reasoning, context provides meaning.*

Thus in quantitative reasoning, context plays a dual role: in its formal mathematical aspects, context obscures structure. But in its interpretive aspects, context provides meaning. This fundamental conflict, I believe, is largely responsible for what gives quantitative reasoning its intellectual vitality, but it is also what makes it so hard to define in operational terms. Like a branch of mathematics, the framework for quantitative reasoning can be regarded as an abstract deductive structure, a triumph of *esprit de géométrie*. But like textual exegesis or historical reconstruction, quantitative reasoning must also be understood as an interpretive activity, animated and guided by *esprit de finesse*.

Thus we may characterize quantitative reasoning as an interpretive activity that takes place within a deductively structured framework. Quantitative reasoning creates a tapestry of meaning that depends for its integrity on the coordination of warp and weft. One set of threads is provided by the abstract patterns; these provide the structural support. But the richness and relevance of the picture depend on how the threads of those abstract patterns interweave with the complementary threads of context and its story line.

A Cognitive Emulsion

The tension between the logical/deductive and the verbal/interpretive is so central, I believe, as to serve as a defining characterization of quantitative reasoning. (It is also present in the sciences and social sciences, of course, but precisely to the extent that these subjects rely on quantitative reasoning.) But the logical/deductive and verbal/interpretive are but two

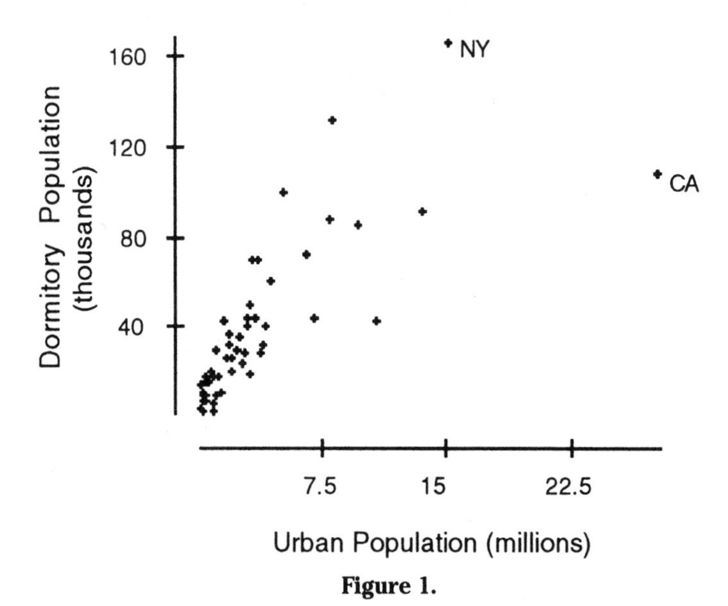

Figure 1.

of four fundamentally different kinds of thinking that form the cognitive emulsion of quantitative reasoning:

- Computational/Algorithmic
- Logical/Deductive
- Visual/Dynamic
- Verbal/Interpretive

An example—a small thought experiment—will serve to illustrate these four different kinds of thinking. Each point in Figure 1 represents one of the 50 U.S. states, with horizontal coordinates equal to the state's urban population and vertical coordinates equal to the number of the state's college students housed in dormitories. It is easy to see just from the distribution of points that states with higher urban populations tend to house larger numbers of students in dormitories: there is a strong pattern of association between numbers of people living in cities and numbers of students living in dormitories.

The correlation coefficient measures the strength of the (linear) relationship between the vertical and horizontal coordinates, with 0 indicating no linear relationship, and 1 indicating a perfect straight-line relationship. It can be estimated by applying the following steps:

1. *Balloon.* Draw a symmetric oval balloon that encloses roughly 90 percent of the points, one that could serve as a summary of the overall cloud of points.

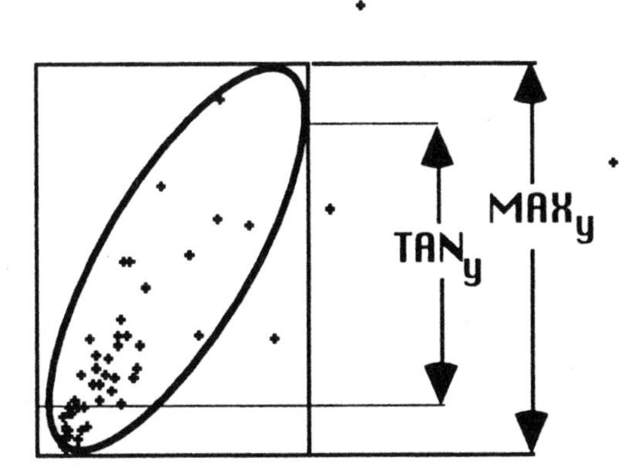

Figure 2.
Estimating correlation using the balloon rule.

2. *Box of tangents.* Enclose the balloon in a rectangle by drawing the two vertical and two horizontal tangent lines.
3. *TANy.* Measure the vertical distance between the two points of vertical tangency.
4. *MAXy.* Measure the vertical distance between the two points of horizontal tangency.
5. *Correlation.* Compute CORREL = ± TANy / MAXy. Use plus (+) if the balloon tilts up and to the right; otherwise use minus (–) (see Figure 2).

This example is a typical blend of story (dormitory versus urban populations), picture (the scatter plot), theory (of correlation coefficients), and rules (the five steps).

Notice how the more attention you pay to any one element—picture, story, theory, or rules—the less attention goes to the other three. In Step 1, to draw the balloon you have to focus on the shape of the cloud of points, and the more effectively you do that, the more the story fades into the background. The shape of the cloud forces your balloon to be something of a compromise: the rule calls for a symmetric oval, but the points are thickly clustered in the lower left, and fan out toward the upper right, where they are sparsely distributed over a comparatively broader range. As you struggle with how best to draw a balloon to summarize this not-so-balloon-shaped cluster, any attention to states, dormitories, and cities will take your attention away from the task at hand. At least in Step 1, though, your focus is necessarily on the points that represent the actual data. In Steps 2 to 4, as you draw the four tangent lines and measure the two distances, your attention goes to the balloon summary, rather than to the data it summarizes. Finally, the purely computational Step 5 arrives, and arithmetic displaces even the visual summary.

To illustrate these various forms of thinking in greater detail, consider these four questions:

- What is the numerical value of the correlation for the data in the example?
- Why is correlation a reasonable measure of association?
- How appropriate is the correlation as a summary of the data in the example?
- Do urban states tend to house more students in dormitories?

Each question can best be answered using one of the four kinds of thinking.

Computational/Algorithmic Thinking

What is the numerical value of the correlation for the data in the example? The usual textbook formula for the correlation is

$$r = \frac{\Sigma(x_i - \bar{x})(y_i - \bar{y})}{\sqrt{\Sigma(x_i - \bar{x})^2}\sqrt{\Sigma(y_i - \bar{y})^2}}$$

If you use this formula, you get a correlation of 0.77. My balloon summary is shown in Figure 2. I find TANy = 33 mm, MAXy = 44 mm, and CORREL = 0.75. If you didn't actually try this yourself earlier, I urge you to do it now, so that you can experience first-hand the mind-emptying purity of the ritual (much as in Eugene Zamiatin's futurist novel *We*—which anticipated Orwell's *1984* by a quarter century—whose main character performs numerical computations with the explicit goal of emptying his mind, almost as a form of meditation). Nothing concentrates the mind quite like algorithmic ritual, and for a mind so concentrated, there is no room for context, for deductive logic, or for visual pattern searching.

Logical/Deductive Thinking

Why is correlation a reasonable measure of association? This is an abstract question whose various answers depend mainly on mathematical thinking. I have listed below, in order of increasing technical sophistication, a set of justifications for the correlation as a reasonable measure of strength of association. Note that in these justifications, context is irrelevant, a nuisance. Algorithmic thinking is absent. Simple shapes are present as objects of thought, but visual pattern recognition is not much in evidence:

(a) *Shape.* The ratio CORREL = TAN/MAX characterizes the fatness of the balloon, ranging from 0 for a perfect circle to 1 as the balloon narrows to a line (Figure 3).

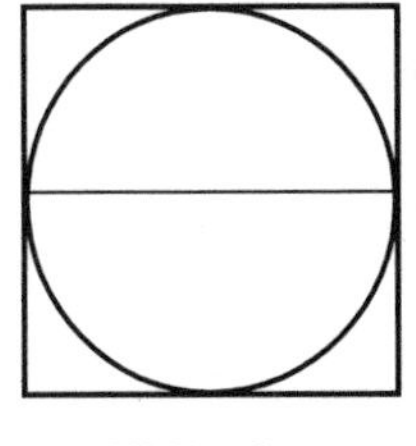

TAN = 0
CORREL = TAN/MAX = 0

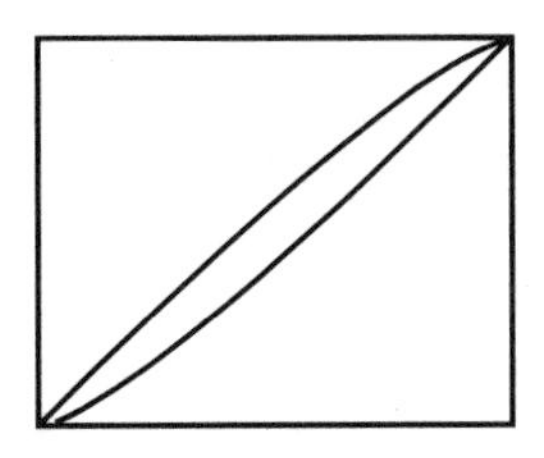

TAN = MAX
CORREL = TAN/MAX = 1

Figure 3.
Correlation ranges from 0 to 1.

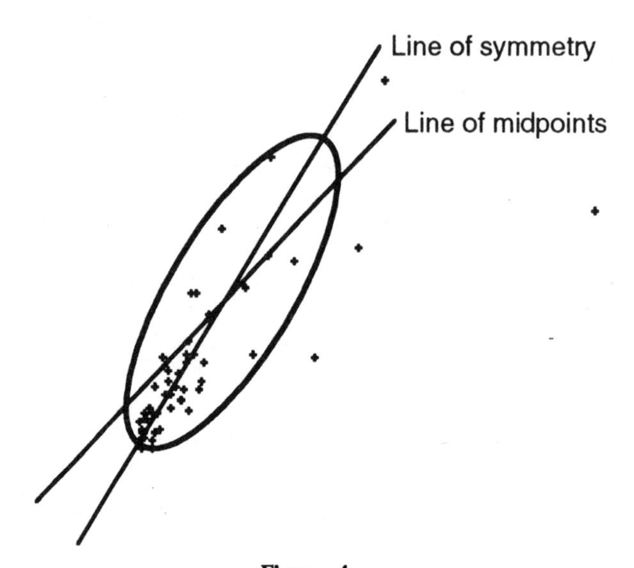

Figure 4.
Correlation measures the size of the regression effect.

(b) *Symmetry.* The correlation will not change if you reverse the axes. Put differently, it is a property of an ellipse that the ratio TAN/MAX is the same in both the vertical and horizontal directions.

(c) *Invariance.* The ratio used to measure the correlation does not depend on the scale. The units for the horizontal axis do not appear, and those for the vertical axis cancel when you divide to get the ratio. Thus the correlation will not change if you switch from pounds to grams, inches to meters, Fahrenheit to Celsius, or nanoseconds to millennia.

(d) *Regression.* The eugenicist Sir Francis Galton observed that on average, the sons of taller-than-average fathers were still taller than average, but also on average not as tall as their fathers. He called this phenomenon "regression to the mean." Abstractly, the regression effect is simply a property of any ellipse, namely, that the line joining the points of vertical tangency is less steep than the major axis (Figure 4). (For any fixed x = father's height, the vertical slice though the balloon at that x value represents the set of corresponding values of y = son's height, and the midpoint of the slice corresponds to their average. The midpoints of all these slices lie along a line, called the regression line, that joins the two points of vertical tangency. Unless the ellipse is oriented so that its lines of symmetry are parallel to the coordinate axes, the major axis will be steeper than the regression line. The correlation coefficient is a measure of how far apart the two lines are.)

(e) *Formula.* The connection between the balloon rule and the usual textbook formula given earlier is not obvious, but it can be made rigorous.

Visual/Dynamic Thinking

How appropriate is the correlation as a summary of the cities-and-dorms data? For this particular example, the answer is, "At best, only moderately appropriate." The points of the plot are not scattered symmetrically, so an ellipse can provide only a moderately good summary of the actual shape (see Figure 2). Since the correlation inherits its suitability from the balloon, it is only as good a summary as the ellipse whose fatness it measures. (One advantage of defining correlation using the balloon instead of the more usual formula is that the balloon definition comes with a built-in visual gauge of suitability.) In the kind of visual thinking involved here—comparing two shapes to gauge how well the simpler one "fits" as a summary of the more complicated one—there is no use for context, no need for deductive logic, nothing to compute, and no algorithm to guide the comparison.

Visual diagnostics of this sort have become very common in data analysis. To oversimplify, but only a little, all of data analysis rests on the bedrock dictum that "Observed = Fitted + Residual" (John Tukey), or, more generally, "Data = Pattern + Deviation" (David Moore), at least according to these statisticians. Once you identify a pattern (the "fit" or "model"), you use the deviation from that pattern to assess its adequacy as a summary. More and more in data analysis, both the search for patterns and the assessment of how well they fit are done visually. Although many of the newer computer plots and dynamic graphics (animations of data) are quite sophisticated, the basic idea—that measuring distances turns sets of numbers into sets of points—is as old as Descartes, as accessible as a high school education, and as ubiquitous as CNN and *USA Today.*

Verbal/Interpretive Thinking

What can we conclude about the association? Do states with large urban populations really tend to house more students in dorms? There are two issues here. How strong is the relationship? What might it mean?

How strong is the relationship? As noted above, the cloud of points is asymmetric and does not naturally suggest an ellipse. Judging the lack of fit was a purely visual operation, but we can also ask how the various visual features relate to the context:

- Many points bunched in the lower left: Most states have relatively small urban populations (a couple of million or so) and relatively small dormitory populations as well (under 50,000).

- The spreading fan: Only a few states have very large urban populations or very large dormitory populations. Moreover, the variability from state to state is larger (more space between points) for the states with larger values.
- Direction: The association is positive. Smaller urban populations go with smaller dormitory populations, larger urban populations with larger dormitory populations.
- Strength: The association is very strong. For all but a few states, knowing the size of a state's urban population allows us to predict its dormitory population to within a fairly narrow range.

What might the association mean? One interpretation suggests itself with all the subtlety of a phone solicitor calling at supper time: since states with large urban populations also have large dormitory populations, cities must attract colleges. Plenty of confirming instances come to mind.

As you probably figured out long ago, this interpretation, although it solicits your patronage with such insistence, is wrong. The pattern of deception involved is so common that data analysts have given it a name—the lurking variable. Both the number of people living in cities and the number of students living in dormitories are in fact indirect measures of the number of people living in the state. Since both give indirect measures of the size of the state's population, it is hardly surprising that the two measures show a strong positive association.

Although I have been teaching data analysis to bright college students for more than two decades, it continues to surprise me how almost universally and automatically they give more attention to the wolf alarm of a high correlation than to the possibility that it amounts to little more than a lupine tautology sneaking about in a cloak of stochastic wool. To strip away the disguise and expose the beast underneath, we have to "adjust for the lurking variable": divide urban population by total population to get the percentage of the population that is urban, divide dormitory population by total population to get the percentage living in dormitories, and plot the result (see Figure 5).

Three features of the plot demand our attention. First, it is much more nearly elliptical than before, so a correlation is now unambiguously an appropriate way to measure the strength of the association, even though there are a few outlying points. Second, the relationship is weaker. The balloon that fits the adjusted data is clearly fatter than whatever skinny compromise balloon you impose on the unadjusted data. Third, the direction is reversed. Urban states—those with a higher percentage of residents living in metropolitan areas—have a smaller percentage of residents living in college dormitories. On reflection, this makes sense. Think about Pullman, Washington, or Ames, Iowa; about Norman, Oklahoma, or Lawrence, Kansas.

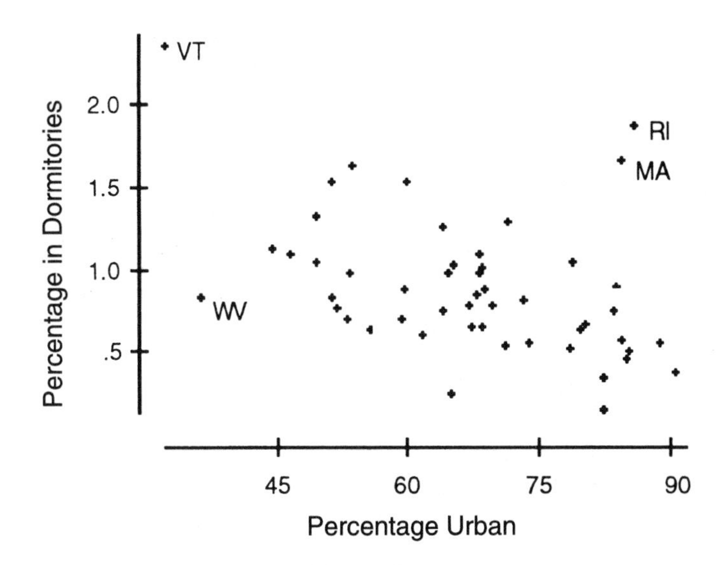

Figure 5.
After adjusting for population, the relationship is less strong, with the direction reversed.

Rural states may have fewer colleges and universities in absolute numbers, but their students make up a higher percentage of the total population of the state. Moreover, think about where those students will be living. Essentially there are two choices: dormitories or privately owned apartments. In relation to the number of students needing housing, do we expect relatively more apartments in Boston, New York, Chicago, and Los Angeles or in Pullman, Ames, Norman, and Lawrence?

Finally, to wrap up the example, notice that in finding the meaning of the observed relationship, the most important element of our thinking was not mathematical, not algorithmic, not visual. It was verbal/interpretive—our recognition that both of the original variables were indirect measures of the same thing, and that we needed to strip away that shared component in order to see more clearly what meaning lay underneath.

A Continuum of Quantitative Reasoning

We are now ready to think of quantitative reasoning along a rough continuum, from the most basic, low-end skills—Can you count? Can you calculate?—to the kinds of reasoning we used in the cities-and-dorms example.

At best, this continuum is a useful metaphor, cruder than many, and certainly not the only approach to defining quantitative reasoning. But it does provide a useful sense of direction while we think about the effect of computers on the past and future of quantitative reasoning.

The low end: calculating in a fixed and familiar context. Everyday examples include making change, telling time or temperature, reckoning dates on a calendar, and measuring—volumes and weights for cooking, lengths for sewing and carpentry. Jobs based on these skills include clerking in a store, operating a saw in a lumberyard, and simple recordkeeping of the sort needed for quality control of a factory machine. The kind of thinking required is mainly computational, in a fixed, repetitive context that makes interpretation an automatic ritual. Variables tend to occur one at a time, so that relationships among variables play little or no role in low-end quantitative thinking. The relevance of these skills to a person's life outside the job tends to be narrow and limited.

The middle: solving problems in a particular applied context. Examples include planning and building as in construction, keeping accounts and handling household budgets, technical benchwork in a laboratory, etc. The context, though circumscribed, is broader than at the low end, and computational work is at times integrated with visual and interpretive thinking. Variables tend to occur in relationships with other variables, although the relationships tend to be well-known and explicit. Quantitative reasoning of this sort often involves a kind of translation—recognizing the structure of a problem, translating to a familiar abstract model stripped of context, finding a solution (often based mainly on algorithmic thinking), and finally translating the result back to the applied context. Here, as with low-end thinking, the relevance to a person's life outside the job is limited, because the areas of application tend to be quite specialized.

The high end: reasoning about relationships. The cities-and-dorms example is fairly typical of high-end quantitative reasoning, which I take to be concerned mainly with reasoning about relationships in numerical data. The contexts tend to be broad and open-ended, the relationships may be unclear and uncertain, and the connections among kinds of thinking are flexible and do not necessarily follow fixed patterns. This kind of thinking qualifies as high end in much the same way that Plato's curriculum put mathematics below (prior to) other, more worldly subjects, because mathematics, being "cleaner," is easier. Just as in Plato's curriculum, it is through high-end reasoning that the pure structures of mathematics make contact with real life and find broad relevance in both civic discourse and personal decision making.

Instances of such reasoning are often more naturally framed as questions to answer rather than as problems to solve:

- What are the likely consequences of the continuing shift toward managed health care?
- Is there a link between income inequality and the practice of financing public education through local property taxes?
- How much of our current national debt is due to the supply-side tax cuts of the 1980s?
- Is capital punishment a deterrent to murder?
- Do cell phones cause cancer?
- Should middle-age men take aspirin to reduce the chance of heart attacks?

These are questions of public policy or personal choice for which empirical data are available and relevant. They require reasoning about uncertain (or unknown) relationships in an applied context whose boundaries are not well-defined. Inevitably, then, just like the cities-and-dorms example, they involve all four kinds of thinking involved in quantitative reasoning.

(For simplicity's sake, I have ignored a possible second dimension, which I think of as measuring a mix of technical sophistication and mathematical ingenuity of the sort found in professionals in fields that use mathematics. This alternative dimension would lead to a different notion of the high end of quantitative reasoning. Examples of this kind of high-end thinking occur in architecture, engineering, economics, and so on. Although this kind of thinking is certainly an important dimension of quantitative reasoning, it requires years of specialized training and thus is not within the reach of a well-educated high school graduate.)

Cheaper Computing, Cheapened Computation

At long last we have come full circle, to ask again: Why the striking asymmetry in the attention paid to verbal and quantitative literacy? Why has quantitative literacy only recently come to be seen as important? Why not sooner?

Mathematics is not the answer. Depending on how much algebra you think should go into a definition of "mathematically sophisticated," we have had sophisticated mathematical thinking for somewhere between three centuries and two millennia. Sophisticated technological applications date back well into the last century. No development in mathematics in the last two or three decades can explain why quantitative literacy has only now come to be seen as important.

Statistics is not the answer, either, even though statistics—as the science of producing and interpreting data—is closely related to quantitative reasoning and has experienced explosive growth during the last two decades, in close parallel with the growth in attention to quantitative reasoning. For example, in 1965, U.S. two-year colleges on average offered only one section of statistics for every ten sections of calculus; by 1990, there were five. The American Statistical Association, two-thirds of whose members come from outside higher education, grew from about 2,000 members in the mid-1970s to nearly 10 times that by the mid-1990s.

The timing of this growth quite clearly coincides with the growth of attention to quantitative literacy. However, it was half a century earlier, during the first few decades of this century, that statistical thinking established its importance in psychology, agriculture, medicine, and industry. By the 1940s and 1950s, universities were teaching statistics courses and even establishing graduate departments of statistics. By the mid-1960s, statistics was regularly taught at a number of two-year colleges, and several thoughtful and accessible accounts of statistics for general readers had been written. See *Design for Decision; Statistics: A New Approach; Facts from Figures;* and *Statistics: A Guide to the Unknown.*[6] And yet it was not until the 1980s that U.S. colleges began developing courses in quantitative reasoning, and not until the mid-1990s that the OECD published its survey of quantitative literacy rates. Why not earlier?

A tempting answer is that attention to quantitative reasoning grew in response to increased use of data in the media. Weather reports that at one time simply told us it would rain turned stochastic and started telling us the chance of rain instead; public opinion polls that once just predicted the margin of an election victory began reporting the margin of error as well; and advertisers who were required to back up their claims with data started reporting some of their data to enhance their credibility. Surely these changes contributed at least a little to the interest in quantitative reasoning, but as an explanation for a major new emphasis, I find them unsatisfactory. Weather reports, opinion polls and advertising claims are often barely more than "data as entertainment," little sugar-coated number-snacks for channel surfers. These trends are just not substantive enough, in themselves, to support such a big change.

It seems all but certain that the three patterns of growth—in statistics, in the reporting of data by the media, and in the attention to quantitative literacy—are linked to each other through one other big change of the last two decades. That change, in the use of computers, is so nearly a cliché that I will not try to present a supporting argument. Suffice it to say that once those little boxes of chips began to pop up like dandelions in one field after another, it was just a matter of time before the air was thick with floating wisps of data. The information in the "information age" is largely *quantitative* information.

The irony, easily overlooked, is that computers have had very different effects at the two ends of the quantitative literacy continuum. While computers have increased the value of reasoning with data, they have simultaneously undermined the value of low-end quantitative skills. As computers have made data omnipresent, cheaper computing has cheapened computation. The availability of ever less expensive, ever more powerful machines has steadily eroded the value of teaching and learning numerical recipes. Examples are so numerous that it is enough to mention only one: running a checkout register at a supermarket. What used to require a fair amount of quantitative skill by a human being is now done by bar codes and scanners; machines even tell how much change to give back. Soon only those elements of quantitative literacy that cannot be reduced to recipes will continue to have value.

Impact on Civic Life

The consequences for the future, although uncertain, are clearly important. Two trends do seem certain to continue: the universe of what computers can do will continue to expand, just as it has ever since the big bang that gave birth to the semiconductor industry. And any work that at one time was regarded and rewarded as skilled but that gets overtaken by the growing range and reach of computing will lose value and sink toward the wage floor of the economy.

Almost as surely, these trends pose two threats to civic life in the United States—the threats of greater economic inequality and of greater political inequality. The economic threat has been addressed in a number of books and articles in recent years, and so I will not go into it here. But the political threat is every bit as important. More and more, debates about issues of public policy such as managed health care and supply-side tax cuts tend to involve quantitative reasoning. More and more, this tends to put such issues beyond the intellectual reach of citizens who have depended on public education. More and more, this fosters a sense of futility, frustration, and cynicism about political leaders, about government, and about the democratic political process.

What response can we make to these threats? Many, like former U.S. Secretary of Labor Robert Reich, call for a greater emphasis on science and mathematics in public education. But I worry that even spectacular success in raising the technical skill level of high school graduates may not address the right issues.

Fantasize for a moment. Imagine that somehow, over the next 10 years, we could magically transform public education so that every high school graduate would know as much mathematics as today's undergraduate mathematics major does at the end of the sophomore year of college. The benefits of even such an unrealistic transformation would surely be disappointing. During the decade allotted for this minor miracle to occur, computers will have become far more powerful, so that even widespread competence in college-level mathematics would offer little protection against the threats of economic and political inequality. We simply cannot hope to outrace the cheetahs at Intel and Microsoft with public education's team of house cats, no matter how energetically we try to direct them.

An effective response, if one is to be found, will not come from trying to stay ahead of computers but rather from redirecting public education to emphasize those kinds of thinking that people can still do better than machines. Such thinking need not be any more technical or mathematically sophisticated than we can reasonably expect of high school graduates. Rather, it depends heavily on flexible integration of the four kinds of thinking involved in quantitative reasoning, on thinking about uncertain relationships among variables, and especially on thinking that emphasizes interpretation in an applied context. In short, we should redirect public education at all levels to put less emphasis on technical and purely mathematical skills and more emphasis on high-end quantitative reasoning.

Endnotes

1. World Bank, *World Development Report: Workers in an Integrating World* (New York: Oxford University Press, 1995).
2. United Nations Development Programme, *Human Development Report* (New York: Oxford University Press, 1996).
3. Organization for Economic Cooperation and Development, *Literacy, Economy, and Society* (Ottawa, Canada: Statistics Canada, 1995).
4. Jacques Barzun, *The House of Intellect* (New York: Harper and Row, 1959).
5. David S. Moore, "Teaching Statistics as a Respectable Subject," in *Statistics for the Twenty-First Century,* eds. Florence Gordon and Sheldon Gordon, MAA Notes no. 26 (Washington, D.C.: Mathematical Association of America, 1992).
6. Irwin D. J. Bross, *Design for Decision* (New York: The Macmillan Company, 1953); M. J. Moroney, *Facts from Figures* (Harmondsworth, Middlesex, England: Penguin Books, Ltd., 1961); Harry V. Roberts and W. Allen Wallis, *Statistics: A New Approach* (Glencoe, Ill.: Free Press, 1956); Judith M. Tanur, ed., *Statistics: A Guide to the Unknown* (San Francisco: Holden Day, Inc., 1972).

Solving Problems in the Real World

HENRY O. POLLAK
Teachers College, Columbia University

Instead of "Here's a problem, solve it," in real life it's "Here's a situation, think about it." Real-world problems are open-ended, with solutions that continually evolve to meet the needs of two masters: mathematics and the external situation.

It has long been recognized that problems are a major component of mathematics, and therefore that learning to solve problems is a major component of mathematics education. Problem solving is also at the heart of quantitative literacy—the use of mathematics in everyday life, on the job, and as an intelligent citizen. Real-world problem solving involves not only mathematics but also some problematic situation outside of mathematics or some real-world difficulty crying out for systematic understanding.

Not surprisingly, there is a large literature on problem solving. What is surprising, however, and rather disappointing, is that this literature reveals very little agreement on what the phrase "problem solving" actually means. Alan Schoenfeld[1] summarizes "two poles of meaning," which he observes are illustrated in Webster's definitions of the term "problem": "Definition 1: In mathematics, anything required to be done, or requiring the doing of something. Definition 2: A question . . . that is perplexing or difficult." Thus, problem solving sometimes involves repeated routine exercises to achieve fluency with a particular technique. This is in accord with the first definition. The other definition sees problem solving as the solution of what Schoenfeld calls "problems that are problematic." In this sense, solving problems is not routine drill, but is at the heart of mathematics itself.

What fundamental idea do these two definitions have in common? What features are shared by a grade-schooler's long-multiplication problem, a question about centroids in a calculus course, and a professional mathematician's settlement of a 20-year-old conjecture—other than that it is common usage to call all three "solving" a problem? A sequence of static definitions of problem solving, expanding upon the two cited from Webster, just will not do the trick. The situations in the primary school, in college, and in the professional life of a mathematician are simply too different for a single static definition.

But they do have something in common: the reduction of something you *don't* already know how to do to something you *do* already know how to do. This definition is dynamic, not static. What you already know how to do varies enormously as students progress in school. The grade-schooler knows addition and single-digit multiplication facts, to which the previously unsolved "problem" of long multiplication can be transformed. The calculus student examines the notion of centroids, and ultimately reformulates the question in terms of certain integrals that are by now familiar. (Of course at some previous point, the definitions of the integrals themselves, and carrying out the integrations, were in their turn the subject of problem solving.) Again, when the 20-year-old conjecture is transformed by a series of ingenious insights into previously known mathematics, the problem is said to have been "solved."

Common to all these examples is that once the problem has been "transformed"—the mathematician likes to say "reduced"—to a formulation in terms of known mathematics, the problem solving is in fact over. What follows, the carrying out of previously learned techniques or application of already known mathematics, is no longer problem solving.

Problem Solving as Problem Transformation

We define a "problem" in mathematics as something you do not as yet know how to do. "Problem solving" is the series of steps that transforms a problem into something you do know how to do. Thus the same question may be a problem for one person, but not for another; it may be a problem for a person now, but not at a later time.

This view of problem solving is far from new. In the context of mathematics education at the elementary level, for example, this point is beautifully made by Don Lichtenberg in the following example:[2] "An auditorium contains 23 rows of 25 seats each and 1 row of 19 seats. How many seats are in the auditorium?" Lichtenberg argues that when a student realizes that the sought-for solution is to be found in the formulation $(23 \times 25) + 19$, the problem solving is over. Finding the standard name for this formulation, namely 594, is not problem solving; it is just the practice of something familiar. This insight is important in many contexts, not the least of which is the continuing discussion of the place of calculators in the elementary school. As Lichtenberg points out, calculators don't solve problems. Calculators find solutions only after people have done the problem solving.

Let us not misunderstand this example. We are not saying that the formulation $(23 \times 25) + 19$ is the only thing that is important or that the student should not be expected to do anything more. The student may indeed be asked to find the standard name of the exact answer, namely 594. Or the teacher may ask students to estimate, to the nearest hundred, the capacity of the auditorium. We are now looking, for example, for the observation that $(23 \times 25) + 19$ is almost, but not quite, $(23 \times 25) + 25$, which is 24×25, or $(6 \times 4) \times 25 = 6 \times (4 \times 25) = 6 \times 100 = 600$. In this case, the problem solving now involves mental estimation strategies that make use of the distributive and associative properties of multiplication.

Estimating (or calculating) the capacity of an auditorium should be part of the problem-solving repertoire of anyone who is quantitatively literate. In this broad context of problems arising in everyday life, in intelligent citizenship, and in a great variety of professions, the term problem solving refers not only to mathematics but also to the real-world context in which the problem has arisen. There are two masters to be served now. One is mathematics; the other is the discipline or situation to which mathematics is being applied. The problem is not solved if the mathematics is perfect but too clumsy to be used in the real world, or if it results in unreasonable explanations and predictions. Similarly, the problem is not solved if a ritual has been found that gives reasonable answers in the real world, but nobody can explain why or how it works. Thus our first observation about real-world problem solving is that the problem can be considered solved only if the needs of both mathematics and the real-world situation to which it is being applied are satisfied.

A good way to begin to understand this dual obligation is through an example. No single example can claim to be typical in terms of either the kind of mathematics used or the field of application. On the other hand, the same kind of thinking goes into real-world problem solving, no matter what field of mathematics or which area of application. That statement may seem extreme. Many books on mathematics for the technical trades, for example, seem to consist mostly of formulas to be memorized and then evaluated by substitution of particular values. What conceivable similarity can there be between such activity and the extended quantitative investigation inherent in a complicated economic or scientific inquiry?

Part of the answer is that the technician's substitution of values in a formula is not the whole story. It is only the final step in a longer process of problem solving. What real situation is the technician examining? How was she led to try to use that formula? How does she know that it is both relevant and good enough to use? What piece of the situation will the formula clarify? Put the technician's exercise into the context of the total problem, and the need to serve two masters will be clearly understood.

A Case Study in Real-World Problem Solving

In order to illustrate real-world mathematical problem solving, we shall outline the history of a specific case as it extended over several decades. The problem concerns the pricing of telephone service. Most of the story takes place in the days when AT&T was a regulated monopoly.

In the fall of 1956, a group representing what was then called the Long Lines Division of AT&T and including accountants and economists met with a group from Mathematics Research at (the then-called) Bell Laboratories. They wanted to talk about the pricing of something called Private Line Service. What's that? Imagine you are a company with offices, factories, distribution warehouses, and customer service centers scattered across the country. You have decided that it is good for business if employees at these various locations feel free to talk with each other. So you come to the telephone company to lease a network to connect these specific locations, and no others, during certain parts of the business day. (Notice that this differs from WATS—Wide Area Telephone Service—which allows unlimited outward calls to particular zones to which you have bought service, and from INWATS (i.e., 800 numbers), which allows people in particular zones to call you. With Private Line Service, only specified locations talk, and only to each other.)

The problem, as we said, was pricing this service. The tariff had been negotiated between AT&T and the Federal Communications Commission (FCC) some years earlier. So why did Long Lines need help? To answer this question, here is a list of some of the factors that a tariff would take into account. There may be some charge for equipment, and possibly for switching, but the bulk of the charge will be for transmission. What criteria of economic fairness come to mind?

- The charge should be independent of which calls are actually made. That's the whole point! Neither the customer nor the phone company may want to keep track of the actual calls. Indeed, the purpose of the network is to encourage communication.
- The charge should be independent of the actual lines used. That is the telephone company's choice, not the customer's. For example, if the company routes calls between New York and New Haven through Hartford—or even calls between New York and Chicago through Los Angeles during certain hours of the day—that routing should not be reflected in the price.
- If the network were to have just two stations, the price should depend only on the distance between them.
- Generally speaking, a network among points that are further apart should cost more than one among points that are closer together.

- If the location of one office is moved a little bit, say to a neighboring town, while the rest of the locations stay the same, then the price should change only a little bit.
- There should be a single answer to the question "What will it cost?" If the computation were to be done again tomorrow, the same price should result.
- Given the locations to be connected, the price should be easy to compute.

So far we have been defining the problem in the context of the real-world situation. This initial stage serves the master representing the particular real-world situation being explored. It is sometimes called "problem formulation" or "problem finding," and it comes before the translation into the domain of mathematics, the other master. The entire process of real-world mathematical problem solving is sometimes called "mathematical modeling."

Creating a Mathematical Model

We proceed now to translate the desirable properties of a tariff for private line service into mathematical terms. The first two conditions stipulate that the transmission charge for private line service should depend only on the locations of the stations to be connected, and not on either the usage or the routing. The third says that for two stations, the charge should depend only on the distance between them. The fourth says that the charge should increase as the points move further apart, and the fifth that the charge should be a continuous function of the station locations. We then ask that the charge truly be a function of the station locations, and that this function be easy to compute. (For a process to be called a "function," there must be a uniquely determined output for any input.)

The agreement between AT&T and the FCC, made long before 1956, was that the transmission charge would be based on the shortest geometric network that could be drawn among the stations to be connected (see Fig-

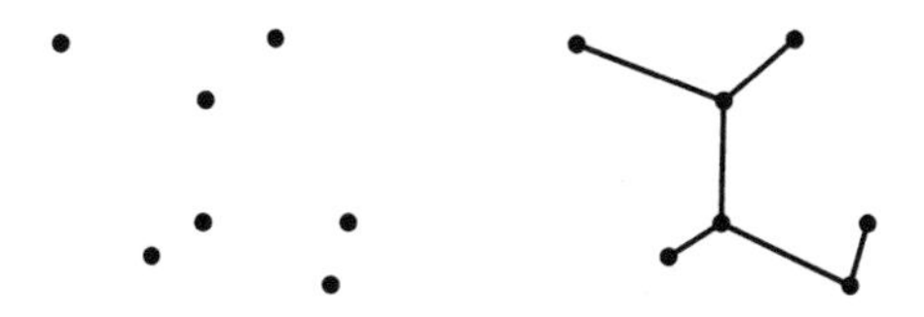

Figure 1.
A shortest connecting network.

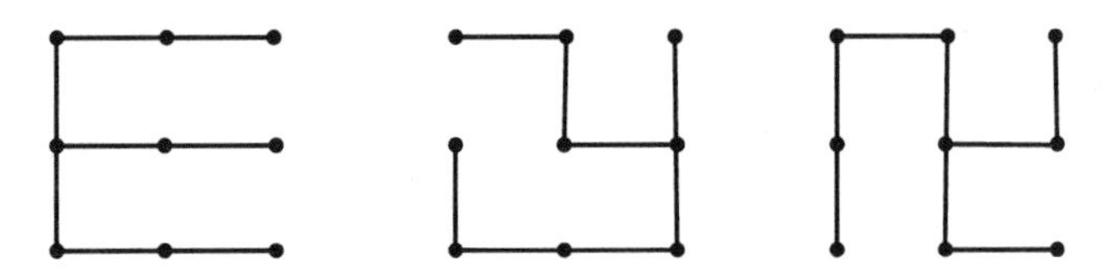

Figure 2.
Shortest networks of equal length.

ure 1). The charge would be so much per mile of length of this mathematical network. It is easy to see that this "mathematical model" satisfies the fairness criteria listed above. The shortest geometric network clearly has nothing to do with actual calls and actual lines. For two points, it is the straight line between them. If the points are further apart, the shortest network will be longer. If a point moves a little bit, the length of the shortest network changes only a little bit. (This requires proving, but is true.) There may be several different shortest networks, but they are all equally long and will therefore cost the same (see Figure 2). Finally, the length of the shortest network should be easy to compute. Easy to compute? But how?

For a number of years, the shortest network had been found by eye on a map at AT&T headquarters in New York. The desire to find a computer algorithm, rather than using a purely visual method, became stronger as business grew. We might also surmise that the technological image of the company would not be improved by photographs of clerks on their knees, with pins and string, poring over a U.S. map on the floor. By about 1955, it had become clear that AT&T needed an algorithm that, given the coordinates of the locations to be connected, would compute the length of the shortest network with these locations as nodes. The largest customer at the time had about 500 stations, and it was urgently desired that this customer's network be computable by algorithm rather than by pins and string.

As I recall, the suppliers of AT&T's main computer had told the two groups attending the meeting that the problem "could not be done." We needed to understand why not. If the desired network has only two nodes, there is only one competitor for a shortest network, namely the straight-line segment between them. If you imagine three nodes, say at the vertices A, B, and C of a triangle, then the shortest network will be the shortest of $AB + BC$, $AC + BC$, and $AB + AC$ (see Figure 3). In other words, you have to consider three possibilities. Another way to understand this is to observe that we can omit one side of the triangle ABC while keeping the other two, and there are three ways to pick the side to be omitted. Why can we omit one side? Because the configuration is a network, i.e., already connected, before we put the last

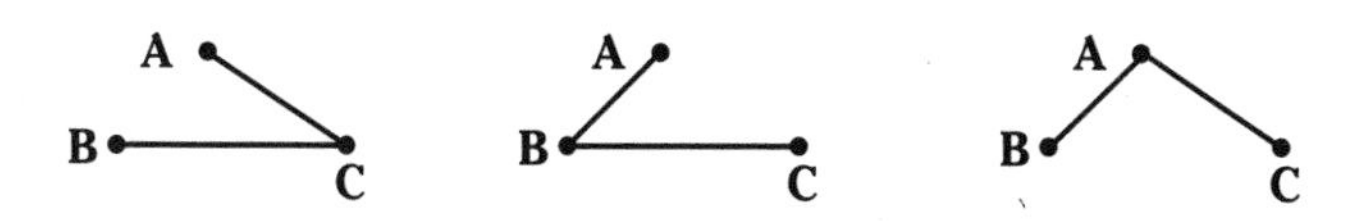

Figure 3.
Three candidates for shortest network among three stations.

side in. So a shortest network will never contain any closed loops. (In mathematical terminology, a configuration consisting of nodes and of edges connecting them is also called a *graph,* and a graph with no loops a *tree.)*

For four points, there are 16 choices. (Can you draw them?) For five, there are 125 (don't try). As a mathematical exercise, what do you think the pattern for the sequence 1, 3, 16, 125 might be? An answer that maintains the numerical pattern is that with n points, there are n to the power (n-2) possible candidates for the honor of being the shortest network. This formula was discovered by the English mathematician Arthur Cayley in the latter part of the nineteenth century.

So one conceivable way to solve the shortest network problem on the computer is to try all n to the power (n-2) different possibilities. How big is n? Well, even with $n = 10$, this would be very unpleasant: one hundred million possibilities. With a design criterion of $n = 500$, forget it!

We needed an easier way to solve the problem. In early 1956, J. B. Kruskal found an algorithm that was much better, but it required more sorting than was practical for computers of that time. So AT&T decided to investigate approximate computer algorithms, and it was for that purpose that they came to visit Bell Laboratories. It soon became clear that AT&T had asked the wrong question. The right question was really not "How do you make an approximation scheme work?" but "Is there some practical algorithm for finding the shortest network?" One of the attendees at the meeting, R. C. Prim, soon solved this problem by producing an algorithm, practical even for computers of the late 1950s, which now bears his name. Everybody was happy, but only for a while.

Unexpected Complications

I was not present during the next stage of this story, so I may not have all the details exactly right. But the central issue is clear. Around 1960, Delta Air Lines had a private line network to connect its three headquarters locations in New York, Chicago, and Atlanta. As luck would have it, these three

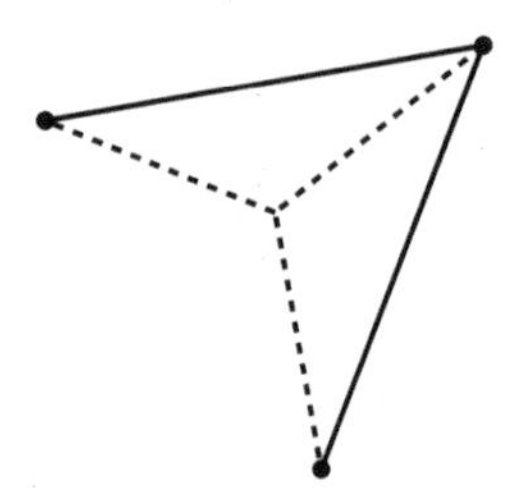

Figure 4.
Adding a new station to shorten the network.

cities lie roughly at the vertices of an equilateral triangle 800 miles on a side. To keep things simple, let us forget the "roughly" and pretend that these distances are exact. The charge for the network would then be based on two of the three sides, or 1,600 miles. But someone at Delta, so the story goes, made the following interesting discovery: If you add to the network a fourth station somewhere in Kentucky—one that you do not need at all—your bill for the private line network will go down! Why? Because you are now allowed to connect each of the three cities to this new Kentucky station (see Figure 4), and the length of this network is about 1,400 miles instead of 1,600. (In an exact equilateral triangle 800 miles on a side, if you put the new point at the center of the triangle, the length will be $800\sqrt{3} = 1{,}385$ miles.) By buying more service than you want or need, you can pay less for it.

This was not a very desirable state of affairs from the telephone company's point of view. What happened was that our list of economic fairness criteria forgot one: If you add a new station to the network, the bill should not go down. Adding this criterion changes the original problem and negates the mathematical model (e.g., charging for the shortest network among the given stations) that was based on it. We needed to compute charges in some other way that satisfied all the previous fairness criteria and also this new one.

What might such a new mathematical model be? The following problem came immediately to mind: Find the shortest network among the given points if you are allowed to add extra internal junctions. So we looked for the shortest network connecting not only the given nodes but also additional internal nodes that would help shorten it further (see Figure 5). A shortest network constructed under these conditions—with whatever additional junctions are needed—will have the property that no new node can shorten it further.

This problem, it turns out, has an extensive history in the mathematical literature. It is named for Jacob Steiner, a Swiss-born mathematician active in Germany in the 1830s, although he did not actually work on it. The earliest reference appears to be to Carl Friedrich Gauss, also in the 1830s. By

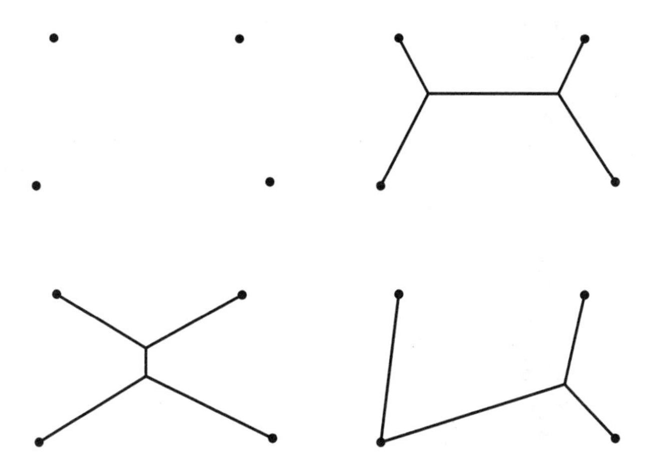

Figure 5.
Different reasonable candidates for shortest networks with additional internal junctions.

1960, quite a bit was known about the problem. For example, if we start with n nodes, we will never have to add more than (n–2) new ones (so Delta was right to add only one unnecessary station). Moreover, at any node that is added, there will be exactly three lines coming into it that will meet at equal angles of 120°.

You might think that all this knowledge would be enough to lead to an algorithm for a practical solution of Steiner's problem, but things are not that simple. In fact, in 1960 no one even knew how to limit the problem to a finite number of steps. The new nodes could be located virtually anywhere. (Cayley's n to the power (n-2) is horrendous, but at least it is finite.) This unfortunate situation was remedied by Z. A. Melzak in 1961, who showed how to solve the Steiner problem in a finite number of steps. Melzak's number was huge—worse even than Cayley's—but at least it was finite. Many mathematicians, including Melzak, some of his students, and a group at Bell Labs, started working on reducing the number to something reasonable (preferably a finite power, like 3, of n). After all, you cannot go to the FCC and ask for a different method of charging for private line service if you do not know how to do the computation.

Considerable progress was made on the Steiner problem for about 15 years. And then we got a shock. In 1977, Michael Garey, Ronald Graham, and David Johnson showed that in a very real sense, there will never be a reasonable algorithm for the Steiner problem. They used tools from the hot new field of algorithmic complexity to show that this problem, and many similar ones, are almost certainly "intractable" in a fundamental sense, namely, that the time required for solution grows exponentially with the size of the problem. So while small problems—those that we could solve as easily with pins and string—can be readily solved, large ones that really need a computer cannot be done with a computer.

So is the story finally over? Almost. Besides the proof of the intractability of the Steiner problem, there was another noteworthy happening in the U.S. telecommunications industry in the 1970s: competition. Other companies started to offer private line service. Only they did not begin by offering it everywhere. One of the earliest competitive services, for example, was between Chicago and St. Louis. This is a very busy route, physically over flat terrain, and it was possible to offer service more cheaply than AT&T did, since AT&T used an algorithm based on distance only, and computed the cost of providing service based on a nationwide average. But in a competitive, as distinct from a monopolistic, environment, pricing based on distance alone, with no regard for the costs of providing the service in the particular situation, is difficult to maintain.

So the problem must change again! A new desirable criterion is that the pricing scheme should be stable under competition. Neither of two competitors using the same technology should have an advantage over the other purely because of the approved pricing algorithm. But then some of the other fairness criteria will have to be modified. So, in fact, the story may not be over, after all.

Problem Solving in the Real World

We can now describe many more characteristics of real-world problem solving. We have alluded to several previously: In real-world problem solving, two masters—namely mathematics and the external situation—have to be satisfied. Many fields of mathematics are important for problem solving in the real world; in fact, probably all the fields ever taught at the school or college level. Furthermore, most areas of human endeavor lead to interesting and important opportunities for problem solving. We could easily get carried away and claim that quantitative literacy *is* problem solving.

In the AT&T example, the real-world problem kept changing, and so did the mathematical problems we were led to try to solve. The interaction between the two masters kept adding to the questions that needed to be examined. Sometimes the evolution was driven by the insights or needs of management, sometimes by those of mathematics itself. These multiple motivations kept the problem exciting—and real. Many fields of mathematics were needed to address the sequence of questions that arose, including geometry, combinatorics, graph theory, theory of algorithms, and complexity theory. None of these look much like traditional applied mathematics. The main area of application was to economics, with some interesting management and legal questions just below the surface. Only recently has it become widely realized that the social sciences are natural environments for mathematical problem solving, much as the physical sciences have been for some time.

The AT&T example was truly open-ended. At the beginning, there was a technical question about a proposed approximation scheme for finding the shortest connecting networks, but this turned out to be the wrong question. The right question was whether there was any practical algorithm for finding those networks. Such a practical algorithm was found, but then Delta showed that the problem was flawed. However, the natural amendment to the problem as formulated led to an infinite problem. Within a year, a way was found to make the problem finite, although admittedly still impractical. Much work followed to try to bring the problem down to size, but by 1977 we knew that it was extremely unlikely that the problem could ever be solved practically. However, by that time the regulatory situation had begun to change drastically, so the problem remained in a state of flux.

Open-endedness is characteristic of many real-world mathematical questions. We often tend to think that questions in mathematics come in only two flavors: "Here's a problem, solve it" and "Here's a theorem, prove it." In the real world, we frequently say "Here's a situation, think about it." Once you determine the problem you want to solve, or the theorem you want to prove, you have made real progress.

A key characteristic of real-world problem solving, certainly visible in the AT&T example, is that progress is driven by considerations of both the external world and mathematics. The motivation for what to do next is a continuing give-and-take between the two. This dimension of mathematical problem solving in the real world reveals perhaps the greatest difference between what happens in the classroom and what happens after graduation. A solution does have to be mathematically correct, but it also has to be practical, it has to give answers that are reasonable in the real-world context, and it cannot have consequences that are unacceptable.

Mathematical Modeling

Let us look a bit more abstractly at the steps in the process of real-world mathematical problem solving. As we said earlier, this process is sometimes called mathematical modeling.

1. We identify something we want to know, do, or understand. The result is a question in the real world.
2. We select "objects" that seem important in the real-world question and study the relations among them. The result is the identification of key concepts.
3. We decide what we will examine and what we will ignore about the objects and their interrelations. You simply cannot take everything into account. The result is an idealized version of the original question.
4. We translate this idealized version into mathematical terms, and obtain a mathematical formulation of it.
5. We identify the field(s) of mathematics that are needed, and bring to bear the instincts and knowledge of those fields.
6. We use mathematical methods and insights, and get results. Out of this step come techniques, interesting examples, solutions, theorems, algorithms.
7. We translate back to the original field and obtain a theory of the idealized question.
8. Now comes the reality check. Do we believe what is being said? Are the results practical, the answers reasonable, the consequences acceptable?
   (a) If yes, the real-world problem solving has been successful, and our next job—both difficult and extraordinarily important—is to communicate with potential users.
   (b) If no, we go back to the beginning. Why are the results impractical, or the answers unreasonable, or the consequences unacceptable? Because the model was not right. We examine what went wrong, try to see what caused it, and start again.

This seemingly ponderous sequence is a lot less terrifying than it looks. What are we really saying? Problem solving consists of seeing something that needs to be done, biting off what looks like a chewable piece, chewing it, and then checking what we have actually eaten. We do that in every field. What is different about real-world mathematical problem solving is that the standards and mental processes of two masters—the real-world situation being studied as well as mathematics—are involved. It is the interplay between the two that is so fascinating. In steps 1 to 3 we are serving the real-world master and trying to formulate the problem to that master's satisfaction. Step 4 represents the transition from one master to the other, and steps 5 and 6 serve the mathematical master; what is done in these must be mathematically sound. Steps 7 and 8 return to the real-world master. Has that master's problem been solved? If not, we return to steps 1 to 3, see where we have gone astray, and the cycle repeats.

Serving Two Masters

Other features of real-world mathematical problem solving are consequences of the intriguing process of trying to serve two masters at once—or, if you prefer, learning to play the two masters off against each other.

Practitioners in the field often invent mathematical techniques that mathematicians find hard to justify. For example, Heaviside's operational calculus involved a "function" that is zero at every point but one, at which it is *so* infinite that the area under this function (i.e., its integral) is 1. It was no use saying that there can be no such function; Heaviside used it in engineering problems, got them right, and taught others to do the same! It took the better part of a century for mathematicians to develop the theory of distributions, which rigorously justified what Heaviside had been doing all along. The fact that really good engineers are almost always right was a major lesson in my professional life. They may not know *why* they are right and their ideas when taken literally may be demonstrably wrong, but they will often turn out to be right in a deeper sense than anyone suspected. Part of the mathematician's job is to find out *why* they are right. Examples of this kind make wonderful stories of real-world problem formulation followed by problem solving.

A closely related feature of looking at the real world mathematically is that many applications of mathematics are best described by "Here's the solution, what was the problem?" Often a practitioner may find a "good" way to do something, but may not know in what sense it really is good. The true question may be to find out what in fact was being *optimized.* In these cases, the real-world master changes the mathematical question to one that the mathematical master would not usually ask: "What optimization criterion leads to the following solution?" This shows that there is more than one purpose to problem solving. It may be insight, or it may be action. Both are worthy causes.

When people talk about a "best" solution for some real-world problem, they often say things like "We will satisfy the largest number of people in the shortest possible time and in the most efficient way." This makes no mathematical sense. You can maximize the number of people served, *or* minimize the time, *or* optimize the efficiency. It is extremely unlikely that among the many possible operating procedures, there will be a single one that simultaneously maximizes coverage, minimizes time, *and* optimizes efficiency. Yet sloppy as the language may be mathematically, the real-world master is trying to say something important: the number of people served, the time it takes to serve them, and the efficiency with which they are served should all be optimal. How can we handle this mathematically?

Let us take a simple example. Consider a bank of elevators in an office building. The manager wants to minimize both the users' waiting time and the owner's bill for electric power. The strategies for accomplishing these two goals will typically be quite different. For example, people will get faster service if you reposition an idle elevator in anticipation of its likely next usage, but that probably does not save electricity. So how do you serve both masters in this kind of problem? What the real-world problem solver does in a case like this is to plot power consumption against expected waiting time, and look for the boundary between regions of possible and impossible points. The choice of boundary point at which to operate is not a mathematical decision, but a managerial one.

Problem solving in this kind of example lays out the possible choices, but does not decide among them. The manager will add other considerations. Why are these not part of real-world mathematical problem solving? In part because power consumption and waiting time are not measured in the same units. If we knew how to measure both in dollars, the problem would be simplified. In some applications, cost is one variable and human lives the other. These are especially troublesome: how do you find a common unit in which to measure costs and lives?

Good real-world problem solving often involves close collaboration among several people. The instincts and insights of two fields are rarely equally resident in one person. Thus joint work is a natural in this area. But this can cause difficulty in an academic setting, where such cooperation is often frowned upon. In business and industry, collaboration comes naturally, both for the reason just cited and to make it more likely that most of the important aspects of a problem will in fact be taken into account. If one person sees 80 percent of what ought to be seen in a problem, then two people (with different backgrounds) working together ought to see 96 percent. (Each misses 20 percent; assuming that the misses are independent, together they miss only 4 percent.) The recent emphasis in education on cooperative learning is thus especially valuable for future employment.

Problem Solving and Quantitative Literacy

The foregoing discussion addresses quantitative literacy in the context of mathematical problem solving in the real world. Real-world problem solving must meet the standards both of mathematics and of the external situation to which mathematics is being applied. This need to serve two masters is the main difference between problem solving in mathematics and mathematical problem solving in the real world. The interplay between the two adds great richness over and above problem solving in mathematics alone.

In addition, the dynamic meanings of "problem" and "problem solving" also become richer. In the real-world situation, as in mathematics itself, the basic meaning of problem solving is the process of reducing something you do not know how to do to something familiar. The collection of the familiar, of what each of us knows how to do, keeps growing over time. This fits well the nature of quantitative literacy: successful mathematical reasoning in a situation outside of mathematics must be both mathematically defensible and useful in the real world, and our mathematical reasoning ability will grow over time as mathematics and external situations change.

A student's mathematics education is simply not complete if that student has not experienced the usefulness of mathematics in the larger world. This experience comes through real-world problem solving. Thus success in mathematics cannot be measured through assessment in mathematics courses alone, or in terms of preparation for the next level of courses. We must also consider the ability to examine in a mathematical way situations in everyday life, on the job, and as citizens. The second master also must be satisfied.

Some mathematicians may find it difficult to accept this point of view. It suggests, after all, some sharing of power with other masters. Real-world problem solving, however, needs to be part of the mathematics curriculum, and cooperative learning part of mathematics pedagogy. Conversations outside traditional mathematics must also enter into the assessment of competence in mathematics. The historically closed system governing mathematics education must change in order truly to accommodate quantitative literacy.

Endnotes

1. Alan Schoenfeld, "Learning to Think Mathematically: Problem Solving, Metacognition, and Sense Making in Mathematics," in *Handbook of Research on Mathematics Teaching and Learning,* ed. Douglas A. Grouws. (New York: Macmillan, 1992), 377.
2. Donovan Lichtenberg, "The Difference Between Problems and Answers," *Arithmetic Teacher* (March 1984).

Quantitative Practices

PETER J. DENNING
George Mason University

It is not literacy but practices that create actions and constitute expertise. Literacy deals with descriptions; practices with data, design, uncertainty, trade-offs. Schools may stress literacy, but many important practices defy description.

Can citizens distinguish the dross from the essential in the musings of technology experts? Can they make sense of newspaper commentaries? Can they understand a risk assessment or tell if it is reasonable? Although not needing the level of quantitative expertise of a scientist or engineer, the average citizen does need to cope with such situations every day.

For three decades, I have been employed as an engineer, working with computing and information technology. Quantitative methods are used extensively in my field. I have also observed them in all the other engineering disciplines, in the sciences, and in many other fields. They are practiced everywhere.

Yet I am concerned that a discussion focused on literacy rather than on practice may not yield the intended educational outcome. In my view, the central question is, "What quantitative practices does a person need to know to be effective as a worker, a homemaker, a citizen?" A focus on literacy can lead to descriptions and observations of practices, but not to the practices themselves. Literacy is like the menu in a restaurant: it tells you about the dinner, but it cannot feed you. The world of practices is messy: practices defy precise description; new practices are constantly emerging; others are becoming obsolete; practices evolve in harmony with technologies. Despite the fuzziness and dynamics of practices, it is essential to understand their importance for what it means to be educated. Then we will be able to draw some new conclusions about "literacy" and about such apparently mundane questions as whether students should be allowed to bring calculators to exams.

A Short Story

A few years ago, my older daughter asked me to help her do her mathematics homework. She was totally stuck on a set of word problems about proportions. (Example: You measure the length of the shadow of a 6-foot vertical stick as 10 feet. You measure the length of the shadow of a tree as 100 feet. How tall is the tree?) She said she understood the concept of proportions but couldn't see how to use it in the word problem. I asked her to explain the concept of proportions. She said: "You're given that *A:B* is the same as *C:D*. The word problem gives you three of the four variables *A, B, C,* and *D*. You plug in the values and solve for the missing value." Sounded impressive. But she was utterly unable to connect this with the word problem. She did not understand how to represent entities with the symbols *A, B, C,* or *D*. I asked her to draw a diagram of the situation behind the problem. She could not do it. She had the same difficulty with the other word problems. She had no conception of how to assign variables or to draw a picture of the situation.

So I said, "Look, I'm going to solve the first five of these problems. I'm going to think out loud and draw pictures. You just watch me do it. Don't try and figure anything out, just watch." By the time I'd finished the fifth problem she said, "I think I see what's going on." And she then went on to solve the other five problems, each one with progressively less assistance from me.

After asking my daughter about her mathematics classes, I concluded that she had never seen a mathematician in action, solving problems. She had never witnessed the *practice* of mathematics. She had never been involved in the practices of assigning variables or drawing pictures. Without these practices, the theory and principles were useless to her. She had been short-changed by her curriculum, which could not deliver what it promised.

I conclude from conversations with others that my daughter was not alone. Many young people cannot practice mathematics after finishing the high school mathematics curriculum. We call that "functional illiteracy" or sometimes "innumeracy."[1] High School Mathematics organizes the principles in a very logical progression, but it does not teach the practice of mathematics. It is as if the designers of the curriculum were stuck in the notion that practice is the application of theory and will follow naturally when a student is well-grounded in theory.

Mathematics is more than that. It is a language, a discourse, and a set of practices. If you do not have a chance to observe a mathematician at work or to work with a mathematician, you will not learn mathematical practices, and it may not occur to you that mathematicians do anything of value. Every time I ask someone to describe how they learned something they seem to do well, they recall a moment at the beginning when they observed people doing it and producing useful results, and they recall later moments when they got involved in doing it themselves. In fact, the practice kindled their interest in learning the theory.

The Importance of Practices

My daughter's story recalls an important distinction between theory and practice. To explore this distinction more deeply, I would like to replace "theory" with the broader term "descriptions." Descriptions are the theories, representations, models, data, facts, rules, and narratives of a domain. A practice is a habitual pattern of action engaged in routinely by people in a domain, usually without thought; practices include the standard patterns, routines, procedures, processes, and habits of people acting in a domain. Mathematicians and journalists operate primarily with descriptions. Managers, sports professionals, and coaches operate primarily with practices. Engineers, scientists, doctors, and lawyers deal with both.

Descriptions and practices overlap, but neither contains the other. Think of the difference between the sports journalist and the basketball player, between the financial analyst and the investor, between the professor of engineering and the licensed professional engineer, or between the menu and the dinner. Journalists can tell us why the ballplayer shoots well, but cannot themselves shoot; ballplayers are notorious for their inability to describe what they do in ways that help others imitate them. The financial analyst tells us why the stock market is rising or falling, but is quiet about his dismal record as an investor. An expert in fluid dynamics describes in detail the method of calculating Euler flows around a wing, but cannot get the algorithm to run fast on a Connection Machine.

We educators incline toward the domain of descriptions. It is our stock-in-trade, the stuff of lectures and presentations. It is where we place all the models and theories of the world that we want our students to learn. We contrast education and training, and locate education in the more familiar territory of descriptions; we harbor suspicions of training, which is about imparting specific practices. We have been brought up on the theory that action happens when we apply a (mental) model of the world to the situation at hand. Descriptions seem rational; practices do not.

Quantitative literacy concerns a student's familiarity with numbers and numerical manipulations. The term "literacy" already reveals a bias toward descriptions. My purpose in the remainder of this article is to suggest that there is a great richness in the practices of working with data and numbers, practices that are not well-captured as descriptions. I suggest that we should examine "quantitative practices" rather than "quantitative literacy" to find the answers to our questions about what to teach our students (see Figure 1).

There are important quantitative practices for which we have no effective descriptions. (By effective, I mean that someone else can take the description, understand it, practice it, and finally appropriate the practice

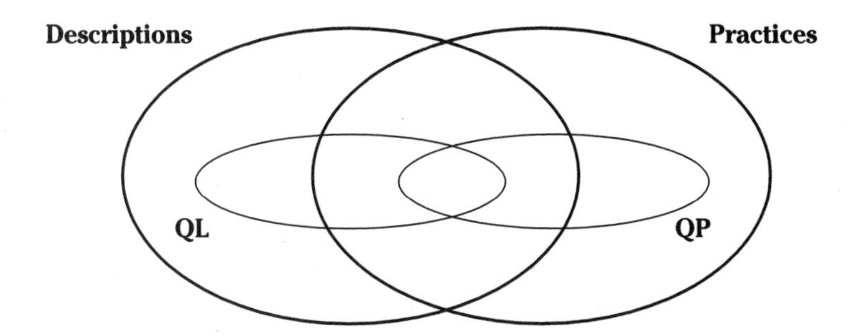

Figure 1.
Quantitative practices (QP) emphasize measurement, evaluation, model validations, and design trade-offs. Quantitative literacy (QL) emphasizes familiarity with descriptions and basic practices. The overlap of QP and QL are the practices for which we have precise descriptions.

for himself or herself.) Here are a few examples of important practices in engineering and science for which no effective descriptions are known:

Fractal art
Formulating a null hypothesis
Designing experiments
Collecting data
Knowing when you have enough data
Approximating very large data sets
Constructing effective reports
Constructing and validating a model
Finding patterns in data
Classifying data into clusters
Estimating and removing measurement errors
Human factors experiments
Designing and validating a heuristic algorithm
Evaluating trade-offs in designing a system

This list can easily be extended to include other domains, for example:

Determining whether a Federal Reserve interest rate hike will induce recession
Modeling a city (e.g., the game SimCity)
Predicting traffic jams during rush hour
Identifying stock market cycles
Estimating when a river will crest after a flooding rain

Calculating the sag of a bridge during rush hour
Locating oil fields from test drillings
Locating fiber-optic links to relieve Internet traffic jams
Routing airplanes among cities and maintenance stops
Determining the throughput of an assembly line
Projecting the financial position of a company for the next two years
Measuring chemical levels in a blood sample
Estimating the pollen count

To the best of my knowledge, most approaches to quantitative literacy assume implicitly that the quantitative practices with which students must be familiar have effective descriptions. For the reasons stated above, I believe this assumption is misleading at best and invalid at worst.

Tools

Practices are usually supported by tools and in many cases are impossible to perform without them. Typing, an important practice in using computers, is an example. Its essential tool is the keyboard, and not just any keyboard: you will have difficulty with a Dvorak keyboard if you were trained on a Qwerty keyboard. As a physical object, a keyboard has no meaning to someone without the practice of typing; it would appear as an utter mystery to a time traveler from the Middle Ages. You are unlikely to succeed at using a computer today without knowing how to type; as you learn computers, you cross-appropriate typing from another domain.

Practices and tools live in a symbiotic relationship: each needs the other, and neither is meaningful without the other. Tools evolve in harmony with practices. The desire for better, faster, cheaper results stimulates technologies that encourage more effective practices. Modern banking, budgeting, and financial reporting are impossible without the spreadsheet. Yet a financial spreadsheet is meaningless to someone who has never seen a list of accounts.

Tools enable practices but do not confer them. The practices themselves are learned from other people. The hand calculator and spreadsheet, which did not exist half a century ago, are essential tools for modern quantitative practice.

Competence

Think of a highly competent person, one whom you might describe as a virtuoso. What do we mean when we say a particular person is competent in a field? We mean that that person understands the history, methods, practices, boundaries, and current problems of that field and its relationships to other fields. More than that, we mean that the person can *perform* effectively in that field. The actions of the highly competent person impress us with their skill and finesse.

The important observation is that we associate competence, effectiveness, and knowledge with actions. We do not expect to be able to perform like the experts simply by listening to their descriptions of their actions or mental processes. Indeed, it is quite usual that the highly competent person cannot give a clear description of *how* he or she performs.

Since the 1960s, the philosopher Hubert Dreyfus has investigated human competence, inspired by the question of whether expert systems (and other machines intended to mimic human behavior) can in principle become as competent as human experts. He identified six competence levels, corresponding to ever higher demonstrated capabilities for performance: beginner, rookie, professional, expert, virtuoso, and master. Dreyfus demonstrated that it takes time for a person to acquire the new skills required to progress to a higher level, and that the highest levels may take many years of practice to attain. His surprising conclusion is that rule-following behavior is not present at the expert, virtuoso, and master levels. Consequently, expert systems could never become "expert."[2] In fact, systems based on rules can hardly be expected to attain the competence level of a professional. In showing this, Dreyfus challenged the conventional wisdom that expert behavior can be described formally with enough precision that a machine could duplicate the behavior or that someone else could learn the same behavior from the description.

Dreyfus's conclusion is important in the present discussion. Many practices are not the "application" of rules and cannot be learned from descriptions. To be "literate" is to be versed in descriptions. The only way to learn the necessary practices is to perform them.

Obsolescence

Some people say that many high school graduates are (quantitatively) illiterate because they cannot "do arithmetic," meaning that they cannot do routine arithmetic calculations by hand, with pencil and paper. These critics say that calculators are crutches that should not be allowed on exams. I find this whole discussion rather muddled. It does not identify the domain (context) in which the calculations are important or the criteria of effectiveness. Where calculations are important, effectiveness is usually correlated with the number of error-free calculations completed. Therefore, a person skilled with a calculator is more effective than someone who does the arithmetic manually, and a person skilled with a spreadsheet is more effective than the one with the calculator. Even the mundane business of purchasing groceries for a large family on a tight budget benefits from a shopper who can do arithmetic with a calculator. The important thing is that the person apply effective practices and have the tools (such as calculators or spreadsheets) to support those practices.

When a practice is no longer effective, we say that it (and its tools) have become obsolete. Thirty years ago, slide rule calculations and drafting were central to engineering. Today the calculator and spreadsheet are central and the slide rule is a historical curiosity; the CAD program has pushed drafting into the dustbins of history. The word processor has made the typewriter obsolete. The speech recognizer is likely to make the practice of typing at a keyboard obsolete within the next generation. The practice of writing letters and memos is being replaced by electronic mail and Internet communication. The important point is that as some practices become obsolete, they are replaced by new practices that make people more effective. The degree of skill expected of the practitioner goes up and the new practices are supported by more sophisticated technology.

Should we worry about students relying so much on calculators that they cannot do their banking or buy their groceries without taking a calculator along? I think not. We should worry instead about students who are not facile with the calculator. Modern shoppers shop for balanced meals within fixed budgets—a more sophisticated practice than arithmetic calculation. Working adults must manage cash flows within budgets and properly report income on their tax forms—also a more sophisticated set of practices than arithmetic calculation. The job of the shopper is greatly facilitated by the calculator, and that of the working adult by the automated checkbook (e.g., Quicken). The person who is competent solely at arithmetic cannot perform at the same level as the person who is competent with the calculator or spreadsheet. And so it may well be that the skill of manual arithmetic will become obsolete.

Quantitative Practices in Practice

Many of the foregoing statements are quite general. I would like to illustrate them with a closer look at several disciplines, beginning with my own, computing. I do not intend to be exhaustive, but I want to drive home the point that quantitative practices are pervasive.

Computing

The discipline of computing has been defined as the body of knowledge about automating step-by-step procedures—i.e., the set of phenomena surrounding computers. It is the discipline whose practitioners help other people take care of their concerns about representing, storing, retrieving, communicating, and processing data, and coordinating their interactions with each other through the exchange of data. The subject matter of the discipline can be represented as a matrix depicting 11 major areas and three processes: theory, experimentation, and design.[3]

	Theory	*Experimentation*	*Design*
Algorithms and data structures			
Programming languages			
Architecture			
Numerical and symbolic computation			
Operating systems and networks			
Software methodology and engineering			
Data bases and information retrieval			
Artificial intelligence and robotics			
Human computer communication			
Computational science			
Organizational informatics			

In each of the empty columns, imagine a detailed description of the kinds of problems addressed, the accomplishments of the discipline, and the open questions. In every case, the processes of experimentation and design rely extensively on quantitative practices. Let me give some examples:

- When selecting an algorithm, a designer needs to know how fast the algorithm will be or how much storage it will require; although many such questions can be answered mathematically, it is increasingly common to answer them with experiments and simulations, especially when the algorithm relies on approximations or heuristics.
- It might seem that designing a computer microchip is mainly an exercise in computer-aided design and logic simulation. In fact, it has turned into a highly quantitative exercise. Designers hesitate to place an instruction in the computer's instruction set unless they can demonstrate that real programs will run faster. Assessing this requires detailed statistical analyses of frequencies of instructions in actual programs.
- Programs such as Mathematica, Maple, or Reduce that manipulate, evaluate, and display mathematical expressions rest heavily on quantitative methods, especially in determining computational errors and in computing the graphs of a function.

- The Internet is a complex web of interconnected computers and network protocols whose managers constantly measure network traffic, manage routing, and reconfigure line capacities to minimize traffic delays and response times. In this work they use sophisticated heuristic algorithms to find near-optimal line capacities.
- Software engineers subject their programs to rigorous tests, seeking to determine whether the proper internal control paths are followed for each possible pattern of input data.
- Designers of learning machines make heavy use of heuristic algorithms for everything from searching for the best next move in a chess game to the Turing test itself ("for how long can a machine fool a human interrogator?").
- Scientific programmers nearly always begin with a mathematical model of the physical phenomenon they wish to study, and then construct software programs for supercomputers to evaluate and display those models. Since the models almost always contain approximations, these programmers must also validate their models against real data.

Other Fields

Engineering. Civil engineers carry out surveys, design experiments to minimize errors in surveying instruments, test structural plans against computer-based models, calculate quantities of materials needed, and estimate costs of construction. Electrical engineers analyze instruction frequencies in program codes, analyze communication channels for errors, and assess the reliability of power and telephone grids. Mechanical engineers build models to compute stresses in structures, estimate the throughput and response time of manufacturing lines, calibrate instruments, machine to ever finer tolerances, and compute the lifts, drags, and turbulences affecting aircraft. Chemical engineers estimate flow and reaction rates in petroleum plants, calculate optimal yields of chemical production processes, compute the properties of materials (such as heat shields) in inaccessible places (such as Jupiter's atmosphere), and compile detailed assays of the chemical composition of unknown substances. Petroleum engineers estimate the yields of oil fields from soils dug up from drill holes.

Astronomy. Astronomers use sophisticated algorithms to detect very faint objects in the visual fields of telescopes. They search for evanescent signals from extraterrestrial intelligences. They calculate whether split images of distant galaxies might be caused by an intervening black-hole lens. They model the evolution of the universe and gather data to support or reject hypotheses about its birth and death. They track local objects such as comets and asteroids and alert the public to their positions.

Environment. Atmospheric scientists monitor rainfall, project periods of drought, track the status of the ozone hole, and compute pollution alerts for urban centers. Meteorologists forecast weather conditions based on current measurements. Oceanographers measure ocean currents and temperatures and use them to project fish movements and conditions (such as El Niño) that will affect weather. Geologists predict earthquake probabilities and calculate flows of toxins through underground waterways.

Bioinformatics. Genetic engineers use computer models to calculate which DNA sequences are most likely to endow an organism with a desired property. They conduct statistical analyses and cross-correlations of DNA sequences recorded in very large data bases. Epidemiologists use computer models to estimate the spread of diseases. AIDS researchers use computer models to estimate the most likely mutations of HIV in preparation for designing new drugs and vaccines.

Medicine. Biomedical engineers seek out and then incorporate rules of thumb, heuristics, and medical guidelines into "intelligent machines" that make good diagnoses. Medical researchers conduct statistical analyses and correlations of medical data, looking for confirmation of hypotheses about the causes or inhibitors of disease. They perform controlled experiments on new drugs or proposed medical procedures. Lab technicians measure blood chemical compositions.

Finance. Financial advisers build spreadsheet models of an individual's assets, income, and expenses, then devise plans for attaining wealth targets for retirement. Accountants compile the cost and revenue projections of a company for several years in advance, and analyze the sensitivity of the results to assumptions used in the forecast. Bankers monitor and calculate cash reserves and the present values of future investments and liabilities.

Economics. Economists build models to forecast national and global economic trends and determine the possible effects of public policies on growth or shortages. They analyze stock market data to look for trends that would interest investors. They calculate the optimal interest rates and money supplies to control inflation.

Management. Management scientists map out organizational coordination processes, measure them, and project whether proposed reorganizations will be effective. They build models of organizations as large feedback systems that can be evaluated to project the long-term effects of current policies.

Law. Lawyers conduct extensive data base searches for court rulings that might set precedents for a current case; these searches often include statistical analyses of the results. They contribute to the development of software that performs routine legal tasks (such as drawing up wills, deeds, and powers of attorney) on home computers.

Literature. Literary scholars analyze texts for the frequency of occurrence of letters, words, and phrases; they use these results for everything from tracking evolving meanings to determining whether a given person might have authored a document (e.g., were the works of Shakespeare really written by Shakespeare? Was Joe Klein really the anonymous author of a political book?). They can then distill guidelines for language usage and develop style checkers and automated thesauruses for desktop computers.

Implications

It is clear that quantitative practices pervade a wide variety of fields. These practices are reflected in the ways people deal with data, measurements, instruments, experiments, evaluations, models, predictions, forecasts, and trade-offs. Quantitative questions arise even in fields such as law, literature, and medicine that traditionally have not been regarded as quantitative disciplines. The computer has created a rich variety of opportunities for people in these disciplines to import quantitative methods and apply them to their daily routines.

Students of the traditionally quantitative disciplines (e.g., science, engineering, mathematics, statistics, and computing) must master the quantitative practices that are important in their fields. Indeed, much of the university curriculum in these disciplines is organized to ensure that students learn these practices. For this reason, college and university faculty are concerned that students enter with experience in basic quantitative practices. These include numeracy, a working knowledge of algebra and calculus, and some exposure to statistics—the practices of gathering and recording data, monitoring errors in measurements, and extrapolating trends.

The foregoing analysis strongly suggests that we need to take a new look at the role of quantitative literacy in education. We need to reframe the question, focusing not on quantitative literacy but on quantitative practices. Much of what looks like "functional illiteracy" is in fact an absence of relevant practices. Curriculum changes intended to eliminate functional illiteracy should engage students in practices; merely offering better descriptions will not help.

Quantitative practices deal to a great extent with numbers, uncertainty, errors in data, design of experiments, creation of models, validations, drawing conclusions from data, making trade-offs, and the like. They cannot be taught at a blackboard. Teachers must involve their students in labs, fieldwork, simulation games, and the like.

In so doing, however, we should be wary of confusing tools (e.g., calculators, computers) with practices. Tools enable and support practices; but people learn practices from other people. Giving tools to students, and even showing them how to use them, is not sufficient to teach practices. Involvement with the practices is the only way to learn.

We must not forget that what constitutes an effective practice is domain dependent. Practices are also time dependent because what is effective today may be obsolete next year. To some educators, this lack of "timelessness" makes practices seem ephemeral and not a worthy part of a curriculum. But practices as a phenomenon are timeless, even if particular practices change over time.

In education and society, we need to grant practices and descriptions equal levels of respect. Many important practices do not have precise descriptions and cannot be learned by listening to and memorizing descriptions. This means we will have to give up our aversion to "training." Training—the learning of important practices—is an important part of education. The masters are skilled performers. Students cannot attain mastery by studying descriptions.

Science, engineering, mathematics, statistics, and computing are pervaded by quantitative practices. These disciplines cannot exist without them, and any young person aspiring to a technical profession needs to know quantitative practices. These days, with the help of the computer, practitioners of other disciplines and professions are finding that knowledge of quantitative methods gives them a competitive edge in their fields. And as citizens, we need to be familiar with basic quantitative practices to cope with life and work in a technological society, to make sense of the data we encounter, and to evaluate risks.

Quantitative practices are the dinners served at the educational table, the morsels described by the menus of quantitative literacy. For life and work, for citizenship and education, students need immersion in the messy world of practice as much as in the packaged world of literacy.

Endnotes

1. John Allen Paulos, *Innumeracy* (New York: Vintage Books, 1988).
2. Hubert Dreyfus, *What Machines Still Can't Do* (Cambridge, Mass.: Massachusetts Institute of Technology Press, 1992).
3. Peter Denning, "The Structure of Computer Science," in *Encyclopedia of Computer Science*, ed. A. Ralston and E. D. Reilly (New York, N. Y.: Van Nostrand Reinhold, 1983). See also Peter Denning, Douglas Comer, David Gries, Michael Mulder, Allen Tucker, Joseph Turner, and Paul Young, "Computing as a Discipline," *ACM Communications* 32:1 (January 1993), 9-23.

From Opportunity to Results

JUDITH EATON
Chancellor, Minnesota State Colleges and Universities

As society moves from education for opportunity to education for results, mathematics builds conceptual skills that are useful both in vocations and in life. As is true of education itself, quantitative literacy is about life, not just employment.

Many college professors claim that today's students are not nearly as well prepared as those in earlier years—especially in basic skills such as reading and mathematics. Is this concern on target? Are today's college students really as unprepared as their professors believe?

Faculty do indicate that students are not as well prepared as they once were. But I do not have any data about the level of preparation of entering students in our system. To the best of my knowledge, the former systems whose merger created our current system did not keep these data. So we are starting from square one on a data base for the new system.

National data show enormous growth in remedial mathematics courses during the past 15 to 20 years, despite evidence that these courses don't work. How should colleges help students compensate for deficiencies in their backgrounds? What is the best way to get entering students up to speed?

I am skeptical about colleges doing this at all. The data I have seen on remedial education generally suggest that higher education is marginally effective in dealing with mild remedial needs and ineffective in dealing with severe remedial needs. Perhaps the best way to get students up to speed is not to admit them until they are ready or almost ready. Given the limited capacity of higher education and the limitations of K–12 education, I believe that we need what I call "transition schools"—learning centers attached neither to a school district nor a college that assist recent high school graduates and returning adults in building their skills before entering college or university. Another alternative would be to redefine our current graded levels of education so that "college" incorporates some of the studies we prefer to associate with high school or middle school.

Don't community colleges exist, at least in part, to provide this kind of transition?

The kinds of remedial needs out there are too severe for community colleges to handle. This is why I suggest transition schools. Further, I believe it to be a poor public policy decision to turn community colleges into remedial centers. They were created to provide democratic access to higher education. And they are part of higher education. Meeting severe remedial needs is not, in my view, consistent with creating a higher education learning community.

In large public university systems, two-thirds or more of mathematics enrollment is in courses that are ordinarily part of the high school curriculum. Should taxpayers pay for students to take these courses over and over again?

I don't mind taxpayers supporting students who try but don't succeed. Your question reflects a major policy shift in thinking about higher education that has taken place since the 1960s. During those plush years we were willing to fund education experimentation such as open admission; we said then, "It's OK to try but not continue if you don't like college." We are no longer willing to do this. We have moved from education for opportunity to education for results.

Students' educational progress is very often blocked by lack of sufficient quantitative skills—sometimes due to lack of effort, but often due to insufficient opportunity to learn. Are you saying that society no longer is willing to give such students a second chance?

Yes, I believe that we are allowing ourselves to move into this public policy position. I am not saying that I like it, but it is happening.

For centuries, mathematics—primarily arithmetic, geometry, and now calculus—has been accorded a position of priority in school and college curricula. Many now argue that knowledge of computers is the new "quantitative literacy." Should spreadsheets substitute for geometry? Should calculators replace arithmetic?

What you are really talking about is a shift in the quantitative skill set we develop in school. There is nothing sacred about arithmetic and geometry.

If our goal is quantitative conceptual ability, I imagine that it can be reached in many ways. There will be lots of quantitative challenges even if we find little use for arithmetic and geometry in the future. "What quantitative skills do you need to function in society?" is the key question to me, not "Are students continuing to develop the quantitative skills we have been valuing for thousands of years?"

Is it possible to establish common standards of quantitative literacy for postsecondary education that would encompass the very different interests and needs of students in vocational and technical programs on the one hand and those in fine arts and humanities on the other?

Yes. One of the major failings of vocational educators (at least in my view) is that they have opted for reduced standards in what I call "generic skills"—mathematics, reading, writing—through courses such as "Business Math'" and "Business English." I am making the old general education argument here, I know. But common standards across all areas are, I believe, as essential to the success of vocational education as to the liberal arts. They are key to ensuring that vocational education is not considered inferior to the arts and humanities. Our problem is that we do not distinguish between liberal arts education as a capacity-building activity as opposed to the liberal arts studies "for themselves" (perpetuating knowledge in the field). If we view liberal arts as the former, they become "vocational." (I probably could not find five vocational educators who agree with me.)

You may have more support than you imagine. The goal of today's vocational education is to provide solid academic preparation in an applied context. That's not so different from the goal of the standards movement—to emphasize high academic quality in contexts meaningful to students. But if you agree that Business Math is not good enough, you must have some notion of what would be acceptable.

I am not sure that I do have support, because I worry that the same words (solid academic preparation) mask some major differences in intellectual expectations. Given that I believe conceptual skills are vocational skills, any mathematics that assists in building those skills would be acceptable. What is unacceptable is a set of standards that stresses the applied at the price of the conceptual, as happens all too often in Business Math. We need both.

College requirements in quantitative literacy seem entirely ad hoc, giving little evidence of a national consensus. Frequently, what colleges require of graduates is less than what they say they expect of entering students—or less than what high schools claim to require of their graduates. What kind of quantitative literacy should colleges require of all their graduates?

I believe that education is about becoming master or mistress of your own fate and, through this, sustaining dignity. I really can't say how this translates into specific quantitative literacy competencies.

Let's be specific for a minute. What method do you use to figure out a 15 percent tip? Do you know anyone who can't do this in his or her head? Does it really matter? Is calculating tips and sales tax a good surrogate for quantitative literacy?

I don't figure out 15 percent. Instead, I take 20 percent and deduct a little (my guess is I run about 18 percent). Hardly precise. Calculating tips and sales tax are at the low end of quantitative ability, so only to that extent are they reasonable surrogates.

Do you know people (other than teachers or scientists) who use algebra? Do you ever use it? What would you say to a son or daughter who resisted taking algebra in school?

No, I don't know people who use algebra. I rarely need it. Nonetheless, to the extent that it helps develop an important way of thinking, it is worth learning. Education is also about acquiring and developing tools for thinking. If algebra is an obsolete or marginalized way of thinking, then we don't need it.

Some people do think algebra may be obsolete. Studies show that employers tend to disregard job applicants' school records, and students persist in the age-old refrain: "Why do I need to study algebra?" Why, indeed, should they study algebra if employers don't care enough to require it?

Because, as I said, algebra develops an important way of thinking. And because education is about more than employment. I am alarmed that we in higher education are so willing to let the employer define us. Work (and it may occupy considerably less of our time in the future) is but one aspect of life. And education, at least for me, is about life.

Organizing Mathematics Education Around Work

GARY HOACHLANDER
President, MPR Associates, Inc.

Knowing how to use mathematics makes significant differences in people's lives. Students deserve to understand the value of mathematics, and can do so if they engage in authentic, rigorous, advanced mathematics in the context of broad career clusters.

"Why do I have to learn this?" is probably the most frequently asked question in U.S. classrooms, and mathematics classes are certainly no exception. While most students generally recognize the usefulness of basic arithmetic, their skepticism about the value of mathematics grows rapidly as they begin to study algebra, geometry, trigonometry, and calculus. Of course, many students never even take these subjects; indeed, the majority of U.S. high school students do not progress very far beyond prealgebra. Even if they are required to complete three or more years of high school mathematics—and that is now the standard in almost all public schools—about half of all high school students get by on a sequence of mathematics courses that pays little or no attention to the topics covered in advanced mathematics.[1]

This state of affairs, which pervades the entire high school curriculum, is now widely recognized, and as widely deplored. For nearly a decade, public commissions, state and local school boards, national associations, teacher groups, researchers, and concerned citizens have been working hard to raise standards, upgrade curricula, and strengthen high school graduation requirements. The mathematics community has been in the vanguard of this effort: it was the first of the academic disciplines to produce comprehensive, specific standards for curriculum and evaluation.[2] It has followed this achievement with a wide range of activities aimed at implementing these standards and improving teaching capacity in the mathematics profession.

Despite this progress, for which the mathematics community can claim significant credit, students still ask why they must learn advanced mathematics. Over the years, we have become much clearer about what we want students to learn. We have even made headway devising better ways to teach mathematics. But when it comes to convincing students that they can put higher-level knowledge and skills in mathematics to good use, we are still struggling. For the most part, the lame responses—"You need it for the next course, or for college." "Can't you just appreciate the intrinsic beauty of mathematics?" or "Trust me."—are the best that we can muster.

In our search for a more credible response, the world of work has been an alluring siren. When the 1989 mathematics standards were announced, the first of four new social goals for education cited as justification for developing new standards was to produce "mathematically literate workers." This led to experiments with a new school subject called applied algebra, which adopts learning-by-doing as the cornerstone of its pedagogical foundation and builds curriculum around problems drawn from electronics, business, the building trades, and other industries.

As it toils to find sufficient numbers of workers qualified to perform today's jobs, the business community has become a strong proponent of higher standards for mathematics, as well as for other disciplines. Recent federal legislation supporting improved preparation for the U.S. work force—such as the School-to-Work Opportunities Act and the Carl Perkins Vocational and Technical Education Act—promotes the world of work as a focus for school improvement. Moreover, key national policies advocate integrating academic and vocational curricula, closely linking classroom-based and work-based learning, and developing industry skill standards that will help inform teachers and curriculum developers about industry's knowledge and skill requirements.

In such a conducive environment, focusing on the world of work to make mathematics more relevant and understandable would seem an obvious and simple path to follow. However, pursuing this path has proved to be quite difficult. With some notable exceptions, mathematics teaching and curricula are still far removed from work. In fact, courses such as applied algebra enjoy an uneven reputation, and many educators would argue that this is much too charitable an assessment. Some vocational educators, especially those teaching more technical subjects such as computer-assisted drafting or advanced manufacturing, have significantly upgraded the mathematical content of their courses. In other vocational classes, however, the mathematics remains at a decidedly low level. Despite nearly a decade of effort to bring about a more integrated curriculum, no more than 2 or 3 percent of today's high school students experience work opportunities that are carefully integrated with classroom instruction.

In short, we are a long way from realizing the potential offered by the world of work to transform mathematics education, especially for those students who have not been able to master much of the subject in traditional mathematics curricula. Understanding why this is so requires close attention to three important areas: the role of work in education, strategies for organizing education around work, and the relationship between education and industry skill standards.

The Role of Work in Education

Americans are quite ambivalent about focusing elementary and secondary education on preparing young people for work. Phyllis Schlafly and the Eagle Forum have crusaded against the School-to-Work Opportunities Act, labeling it a "communist conspiracy" designed to give the federal government control over the production of future workers. Ironically, not long ago such avowedly Marxist reformers as Samuel Bowles and Herbert Gintis attacked many of the same policies as capitalist tools designed to reproduce the inequalities of race and class that serve the agenda of corporate America.[3] Many others, although less ideologically inclined, have viewed vocational education suspiciously, calling it a "dumping ground" or a "dead-end detour" that turns students away from the road to college and a successful career.

While these viewpoints may be extreme, they reflect concerns that in more moderate forms are shared by almost all educators, parents, and the general public. Americans generally believe that while education should help prepare all students for productive adult lives, it should not be the servant of narrow, short-term demands for labor from the business community. Most also agree that it is wrong to forcibly track young people into particular kinds of jobs, especially when such coercion reflects gender, race, or class biases about who can do what in the work world. Encouraging students to pursue career paths rather than training for specific jobs may help to broaden their future options and reduce the potential for this kind of discrimination. However, it is still debatable whether it is wise to counsel high school students to choose any career track until they become aware of all the options.

Focusing on work, therefore, to make mathematics (or any other academic discipline) more engaging and understandable creates a dilemma. On the one hand, reference to the workplace holds the promise of making mathematics concrete, realistic, and lucid. On the other hand, a work-centered approach can unwittingly seduce educators to force young people into making inappropriate, premature decisions about their futures. Resolving this dilemma requires paying careful attention to two aspects of the role of work in education.

First, the primary purpose of using work in mathematics instruction should be to provide a focus to make mathematics more authentic, thereby helping students to master knowledge and skills that are applicable to a wide range of situations. The purpose should not be to prepare students for particular occupations, to increase the likelihood that they will be more efficient solvers of specific work-related problems, or to make them better users of the latest technology or machinery. If some of these other outcomes occur as a result of realizing the primary objective, that is fine, but they should always be viewed as second-order objectives.

Second, to ensure that work-related applications of mathematics reflect breadth and sophistication, it is essential to conceive of work broadly and for the long term. In other words, it is crucial to define the work focus carefully. For example, concentrating on applying mathematics to problems encountered in autobody repair is far more limiting than addressing how mathematics is used throughout the entire automotive industry (including engineering, manufacturing, marketing, distribution, finance, and so on). Similarly, focusing on mathematical requirements for entry-level jobs in an industry rather than on the full range of positions will most likely produce a strong emphasis on simplistic, low-level applications of mathematics.

As an example, consider how some aspects of introductory algebra might be illustrated with applications from the aviation industry. Imagine a mathematics teacher and a class of ninth- or tenth-grade students with a strong interest in flight, airplanes, and aeronautics. (More will be said later on how this confluence of interests might occur in schools.) These boys and girls may or may not be considering careers in aviation. That really does not matter because they do share a fascination with flying. How might this interest be used to help them understand algebra?

As with the design of most things, designing airplanes involves trade-offs. For example, we want airplanes to go fast, but we also want them to be easy and safe to land using reasonably short runways and touching down at slow speeds. How this trade-off is made depends to a large extent on understanding the mathematics of lift. For example, look at the following equation:[4]

$$L = C_L S(\rho/2) V^2$$

where

L = Lift, in pounds,
C_L = Coefficient of lift
S = Wing area, in square feet

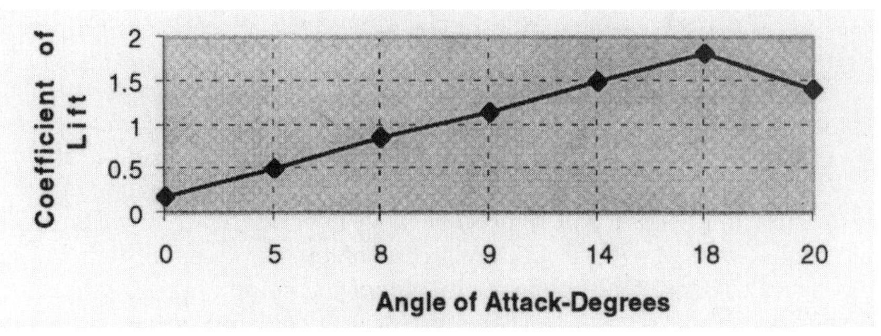

Figure 1.
Coefficient of Lift versus Angle of Attack—NACA 23012 Airfoil.

ρ = Air density, which for standard sea level conditions is 0.002378 slugs per cubic foot, and
V = True velocity (true airspeed) of the air particles around the wing in feet per second.

The coefficient of lift varies depending on the design of the airfoil and the angle at which the airfoil "attacks" the relative wind. An airplane with a high coefficient of lift will land at lower speeds and use less runway, other things being equal, than an airplane with a lower coefficient. However, achieving a high coefficient of lift requires a thicker or more curvilinear airfoil. These properties increase drag, which in turn reduces an aircraft's maximum speed.

Figure 1 shows that the coefficient of lift increases proportionally with the angle of attack, until the angle reaches the point where the airfoil stalls and the coefficient drops off rapidly. Figure 2 shows a similar relationship for a thinner airfoil, such as that used on jets; the maximum coefficient of lift for this airfoil is only half that of the thicker one. So, a plane with the thinner airfoil will fly much faster, but it will also land faster. In fact, it may

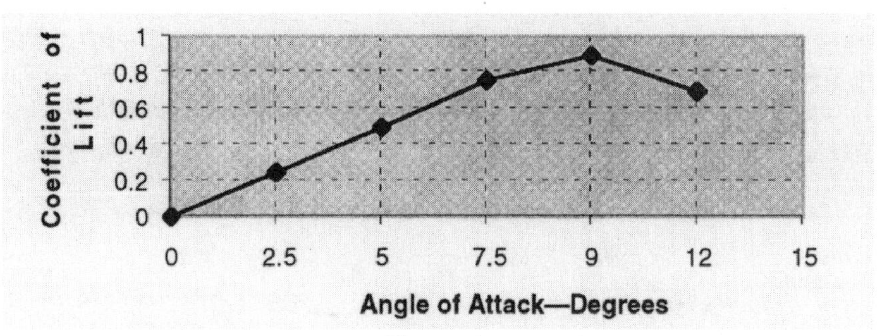

Figure 2.
Coefficient of Lift versus Angle of Attack—NACA 0006 Airfoil.

land so much faster that it cannot land at most airports. This is also why a Piper Cub could land unassisted on an aircraft carrier, but the typical jet fighter requires a tailhook to arrest the landing. The Cub would be much better suited for landing and taking off on the carrier. It would not, however, be very effective in combat!

With this information on the coefficient of lift for these two different airfoils, students can compare the landing speeds of two airplanes alike in every respect except for airfoil design. By assuming that lift must just equal the weight of the aircraft at touchdown, students can use the lift equation to solve for landing velocity using the two maximum values for the coefficients of lift, 1.8 and 0.9, respectively, when they are given appropriate data on wing area and the value of ρ. Furthermore, using this equation, students can experiment with their own airfoil designs, analyzing how changes in either wing area or design affect touchdown speed. Further pursuing this line of inquiry, we can develop the mathematics of flight in considerably more detail and at much higher levels of difficulty, addressing real problems that aircraft designers routinely encounter.

To repeat, the objective of this kind of teaching strategy is not to train students to be aeronautical engineers or airframe and power plant mechanics, although some may choose to pursue such careers. Rather it is simply to use their interest in flight to understand better how mathematics is used by aircraft designers. Students will then begin to see the value of knowing algebra. With this understanding, they will be better motivated to learn algebra, even if it is not always taught in the context of their particular interests. Thus, learning algebra in the context of aeronautics will help students understand basic concepts and skills that they can apply to their career choices no matter what those eventually turn out to be.

Not all students, of course, are interested in flight and aircraft design. Nor, for that matter, are most mathematics teachers. If emphasizing work to engage students' interest in mathematics depends both on identifying their particular work-related interests and on teachers being able to tailor their mathematics instruction to these concerns, can the typical mathematics teacher possibly assume such a role? The answer depends partially on how well curriculum and teaching are structured to facilitate an effective and creative focus on work.

Organizing Education Around Work

Ask most mathematics teachers whether it is desirable to integrate mathematics instruction with real applications in the world of work and the answer will be affirmative. However, press them on whether they do—or even can—

integrate their curriculum and their responses will be much more mixed. In some cases, mathematics teachers simply do not know much about how mathematics is used by people other than mathematicians or mathematics teachers. In other cases, even if these teachers know or are willing to learn something about how mathematics is applied in work settings, the situations are too varied and cover such a broad array of industries and occupations that integration will be shallow and scattered at best. What can a mathematics teacher do to present mathematics to a class of students whose individual interests span subjects as diverse as cars, cosmetics, clothes, computers, space, sculpture, graphic design, and rock music? Although interesting and complicated mathematics applications are embedded in each of these subjects, expecting a teacher to master them all is unrealistic.

There is no easy solution to this problem, but two strategies can help. First, reaching some agreement on how the mathematics curriculum might be organized around work would help both students and teachers identify a work-related domain upon which to focus teaching and learning. Second, schools also need to identify ways for teachers and students to organize around common work-related interests to explore these with some depth and coherence. Thus, reorganizing both the curriculum and the delivery of instruction represents two issues that need careful attention.

Reorganizing Curricula

Traditionally, most high schools have divided the curriculum into academic and vocational subjects, with academic courses constituting about 80 percent of the average high school student's work load and vocational courses the rest. Except for some generic introductory courses such as keyboarding, agriculture, business, or homemaking, the vocational curriculum has consisted of rather narrow courses that direct students toward occupations that require less than a baccalaureate degree. Although this characterization is a bit stereotypical—a small but growing number of high schools offer some powerful alternatives—it is the norm in many schools.[5]

This traditional curricular structure presents many difficulties. At the most basic level, it promotes separation of academic and work-related knowledge. This separation encourages the teaching of academic subjects in an abstract fashion disconnected from context, resulting in courses that are beyond the grasp of many students unless significantly watered down. When work-related knowledge is offered in vocational courses, the domain of work is usually quite restricted, consisting mainly of the trades and prebaccalaureate occupations in agriculture, business, and health. Despite this limited view of work, there are presently between 50 and 400 different vocational "programs," depending on how we count them, including such specialized offerings as small-engine repair, nursing assistant, masonry and drywall, ornamental horticulture, and clerical occupations. Thus, even the most enthusiastic mathematics teacher who supports curricular integration will soon become mired in a bog of unrelated, narrowly confined vocational subjects.

One way out of this morass is to define the world of work much more broadly. For example, Table 1 presents a framework that organizes the work world into 16 major industries. Together, these account for virtually all paid activity in the U.S. economy. Any one of these 16 industries could provide the thematic focus for organizing mathematics instruction (as well as other academic disciplines) around work.[6] For example, rather than concentrating on unfolding the mathematics embedded in auto body repair, auto mechanics, or diesel engine repair, a mathematics teacher could present mathematics in the context of how it is used throughout the entire transportation industry. To be sure, teachers might still want to introduce examples relevant to engine diagnostics and repair, but they could also draw from automotive engineering, safety, marketing, finance, highway planning and construction, and regulatory policy. Linkages between land, sea, and air transportation would also be fair game.

Defining domains of work this broadly has two important advantages. First, it is much more interesting to explore the territory staked out by a major industry than that represented by a relatively limited occupation. Mathematics teachers will find the finance industry as a whole much more fertile for developing applications than accounting or subaccounting clerical occupations. Second, unlike more occupationally specific vocational programs, a broader industry focus does not automatically limit students' options for postsecondary education and employment. Rightly or wrongly, the mere title of a nursing assistant program connotes that it is intended for students who are not likely to pursue a four-year college degree, and perhaps not even an associate's degree. In contrast, a health industry program does not create the same presumption of limited future options.

Table 1.
An Industry-Based Framework for Education for Work.

Agriculture	Health Care
Arts, Culture, and Religion	Hospitality
Built Environment	Insurance
Communication	Manufacturing
Education	Natural Resources
Energy	Personal and Business Services
Finance	Retailing and Wholesaling
Government	Transportation

Source: John Gnaedinger, STS Consultants Ltd.

The broader focus, however, can have disadvantages. One of the most obvious is that its very breadth may encourage instruction focused on rather meaningless generalities or many disconnected ad hoc examples. Furthermore, adopting an industry focus does not mean abandoning all the specific knowledge and "hands-on" learning opportunities offered by traditional vocational courses. Learning how the automobile engine works, for example, can provide students with excellent opportunities for studying and understanding ratios, as well as other mathematical concepts. In fact, exploring the mathematical principles behind the operation of various engines—for example, piston, turbine, and rotary—can make mathematics an exciting and lasting experience for many students.

Nevertheless, 16 industries are still too many for a single teacher to master, and while the number might be reduced by combining some industries into even larger ones, teachers would still face a difficult task. Some new structural strategies, therefore, are required.

Reorganizing Instruction

If teaching and learning are to emphasize mathematical applications in major industries, how can instruction be organized to maximize common interests? Ideally, a particular teacher should not have to plan instruction that simultaneously attempts to satisfy diverse student interests in art, health, finance, transportation, or some other industry. Additionally, individual students should have the opportunity to develop their particular industry interests in some depth, undistracted by other paths of inquiry that may be of little interest. Merely providing a student with work-based examples of mathematics applications is no improvement over traditional instruction if the student has no interest in the context in which mathematics is applied. Indeed, this problem has plagued the implementation of applied mathematics in many high schools. Sadly, there are too many instances of teachers blithely producing example after example of electronics-related mathematics in applied mathematics classes full of students with no interest in electronics.

To help solve this problem, a growing number of high schools are adopting a variety of approaches that allow students and teachers to come together around a common, work-related interest or theme. In its most minimal form, such an approach may consist of nothing more than a single mathematics teacher in a large high school choosing to organize some or all of a one-year course around a particular industry—communications, for example. Students who think this might be a more interesting way to learn algebra or geometry can opt for this teacher's class over the same subject taught by another teacher in a more traditional format. As long as students can choose their teachers, this arrangement allows for some sorting around an industry-based theme. The opportunities for this type of instruction, however, are in reality relatively rare.

At the other extreme, in a large city students might be able to choose among several high schools, each of which is structured around a major industry. Such an arrangement has existed for more than 50 years in New York City, where students may choose among such schools as Aviation High School, the New York High School for the Performing Arts, or Murry Bergtraum High School of Business and Commerce. Some other cities have built magnet schools around a particular industry, for example, the High School for Agricultural Science and Technology in Chicago. In these schools, an industry theme provides the focus for the entire curriculum. Such a curricular structure offers the opportunity to integrate mathematics not only with a particular domain of work but also with other academic subjects.

It is important to note that in these high schools, mathematics does not disappear as a distinct discipline. On the contrary, more often than not there is still a mathematics department, and course titles may not even differ from those in other high schools. However, the context in which mathematics is taught, as well as the way it is taught, usually differs considerably. In an aviation high school, for example, the mathematics problems relate to flight and aircraft, and students learn mathematics concepts and test them in an aeronautics laboratory or on a shop floor where there are real engines and airframes.

Between these two extremes, the single course and an entire school, is a continuum of other strategies. Some high schools organize "schools-within-schools," each with a different industry theme. In a high school of 1,000 students, for example, we might find four academies—health, transportation, finance, and communications—each with its own complement of academic and vocational teachers who share the particular industry focus as a common interest. Other high schools have created new approaches to scheduling. Some divide the school year into three parts: September through December, January, and February to June. The month of January is devoted to student-teacher projects in which students are encouraged to integrate and apply what they have learned in the prior semesters. Still other high schools require an integrated senior project for high school graduation. In some of these schools, the senior project must have a work-based or career theme, while in others students may elect any subject that interests them.

Creating options of this sort is without question more difficult in smaller high schools, especially those located in relatively isolated rural areas. However, even in these schools, the limiting factor is the number of choices, not the ability to organize instruction around a major industry theme. Agriculture is an obvious choice, but in some rural areas natural resources, energy, or even hospitality may be appropriate alternatives. The essential element is choosing a theme that is simultaneously broad enough to interest a large number of students and provide them with opportunities to apply their learning to the work options in their community, while also sufficiently defined to create a clear focus for teaching and learning. There are now enough schools with experience using these kinds of strategies that the standard excuse that it has not yet been tried is no longer credible.[7]

Major industries, then, can provide a useful focus for organizing education around work and for beginning to develop coherent work-related curricula. The adoption of an industry focus, however, still leaves much unspecified about curriculum content and the level of mastery expected of students. To address these issues, attention must be directed to standards, both academic and industrial.

Academic and Industry Standards

The past ten years have seen unprecedented efforts to develop standards in education. Beginning with the mathematics standards in 1989, every major academic discipline has now produced statements about what students are expected to know and be able to do. Some of these statements have been controversial; some are rather vague; and probably none enjoys universal acceptance, either within its profession or among the public at large. Nevertheless, singly and together, they represent an extraordinary determination to clarify what the nation expects students to know as a result of the enormous resources devoted to formal schooling.

With respect to industry skill standards—that is, what a person needs to know to succeed in a particular line of work—there has also been much activity. The U.S. Departments of Education and Labor together supported 22 industry skill standards projects that have reported their results to the public.[8] While much less comprehensive than the various academic standards and probably varying more in both concept and content, these efforts also represent a significant advance in defining the knowledge and skills needed to perform effectively in many of today's work places.

For the most part, these two efforts to specify standards have developed independently of each other. Very few people involved in crafting the academic standards also participated in the development of the industry skill standards and vice versa. Consequently, the academic standards make very few references to the world of work and the industry skill standards, while certainly addressing some academic requirements, tend to adopt a limited and low-level subset of academic knowledge and skills.

For example, the National Automotive Technicians Education Foundation (NATEF) developed *Applied Academic and Workplace Skills for Automobile Technicians.*[9] In many respects, this effort is exemplary, documenting a comprehensive and specific list of academic skills needed by today's automobile technician. In mathematics alone, the NATEF standards list 51 different mathematics skills with examples of how the typical technician applies them on the job. These skills include the following:

- The technician can multiply numbers that include decimals to determine conformance with the manufacturer's specifications;
- The technician can construct a chart, table, or graph that depicts a range of electrical measurements for comparison of sensor and system operational conditions;
- The technician can convert electrical measurement variables presented in one form to another that allows an algebraic solution based on known information (e.g., converting $E=IR$ to $R=E/I$); and
- The technician can determine the volume of a vessel where the specification is in liters.

NATEF's complete list of mathematics-related skills covers addition, subtraction, multiplication, and division; decimals, ratios, and percentages; inequalities; estimation; English and metric measurement; charts, tables, and graphs; measurement of angles; volume, time, and temperature; and some algebra, probability, and statistics. Overall, however, the mathematics described is elementary, with very little content above the eighth-grade level. If a mathematics teacher tried to use the skill standards developed for automobile technicians to convince students of the value of achieving high-level mathematical knowledge, the teacher's credibility would be quickly undermined.

The difficulty here is not with the standards for automobile technicians. There is little doubt that the mathematics abilities described by NATEF are indeed those needed to be a successful technician. The problem is that this effort does not represent a statement of industry standards. Rather, it is simply a statement of occupational standards, and for a very specific occupation at that. Perhaps more precisely, the NATEF document presents an industry standard for automobile technicians; it does not, however, develop a set of standards applicable throughout the industry.

A similar approach was adopted by many of the other industry skill standards projects. Thus the Electronics Industry Foundation focused on knowledge and skills that entry-level electronics technicians need. The health standards project consciously excluded paying any attention to health occupations that are licensed, virtually guaranteeing a focus on low-level jobs. The standards for wholesaling and retailing were limited to requirements for entry-level clerks. Consequently, the mathematics content of these efforts, as well as expectations for other academic disciplines, is relatively low.

It should be stressed that the industry skill standards projects received practically no guidance on how to define "industry" or how to approach the development of standards. Moreover, in some instances, the projects were even encouraged to concentrate on entry-level occupations. Although they did this assignment well, the results fall well short of providing a work-based platform for significant curriculum upgrading or other forms of instructional improvement.

It could be argued that rectifying this situation is simply a matter of developing standards for more occupations, including higher-level ones, across the full range of U.S. industries. However, this strategy is almost certainly doomed to fail. It is difficult to imagine that the nation has either the time or the resources to develop a comprehensive national system of skill standards for every industry defined as finely as photonics or industrial laundering or for every occupation classified as narrowly as automobile technician or retail sales clerk.

From an educational perspective, it would be much more fruitful to develop thoughtful examples of ways in which the full range of academic knowledge and skill, from low to high, are used in a major industry. Even though it is useful to have examples of how automobile technicians use elementary mathematics, it would be even more useful to have illustrations of how more advanced knowledge and skill in algebra, geometry, trigonometry, calculus, probability, and statistics are applied in the automotive world. Such illustrations should be representative of uses throughout an industry. They do not, however, need to be exhaustive, covering every single job classification or subsector of that industry.

Realizing this objective is likely to require a different approach to developing industry skill standards than the one employed thus far. These development efforts need to expand beyond entry-level job requirements and recognize the kinds of knowledge and skills that workers will need to advance in the industry over a lifetime of productive employment. Additionally, academic educators must become more involved in the development process, working with vocational educators and industry representatives to help clarify where and how mathematics at all levels is applied. Industry itself often fails to recognize the knowledge and skills that workers use in their jobs. It is not uncommon, for example, to hear employers claim that most workers have no need for algebra, while at the same time bemoaning the inability of employees to use spreadsheets. Contrary to the conventional wisdom that employers are the best source of information on job-relevant knowledge and skills, professional mathematicians and mathematics teachers may be better observers of where and how people use mathematics in their work. However, to find rich, appropriate applications of mathematics to bring into the classroom, mathematics teachers will need to immerse themselves in an industry outside their own professional domain.

High Standards for the World of Work

What, then, are industry's expectations concerning the quantitative literacy of today's and tomorrow's workers? Ironically, those expectations are much higher than either employers or educators realize, because both have been viewing the world of work much too narrowly. Most advanced high school mathematics has rigorous, interesting applications in the work world. For example, graphic designers routinely use geometry. Carpenters apply the principles of trigonometry in their work, as do surveyors, navigators, and architects. Calculus is the tool of engineers and is widely used to design and shape the built environment, including all modes of transportation (e.g., automobiles, trains, planes). Algebra pervades computing and business modeling, from everyday spreadsheets to sophisticated scheduling systems and financial planning strategies. Statistics is a mainstay for economists, marketing experts, pharmaceutical companies, and political advisers.

In short, the world of work is not at odds with promoting the high mathematics standards sought by mathematics teachers, the higher education community, and the public at large. On the contrary, emphasizing work-related applications can make higher-level mathematics easier to understand and use—not only for students who are less able to learn without concrete explanations and experiences but also for those blessed with the ability to learn in spite of an abstract, disconnected mathematics curriculum. To realize this potential, we do not have to abandon or dilute traditional high school mathematics subjects. Although there may be other, and perhaps even better, ways to structure high school mathematics education, the traditional constructs of algebra, geometry, trigonometry, probability, statistics, and calculus provide powerful approaches to organizing and communicating mathematical knowledge and skill. However, making such knowledge and skill accessible to significantly more students depends on adopting instructional strategies that make the usefulness of mathematics readily apparent. Connecting mathematics education more closely to the world of work offers the promise of benefiting both.

"Why do I have to learn this?" By failing to answer this question, we send students a powerful message that mathematics has little utility and is therefore one more mindless requirement in the drudgery of the high school experience. But knowing how to use mathematics makes significant differences in people's lives. The ability to do mathematics affects whether airplanes fly, a car passes a smog check, a potentially life-saving drug is tested properly, a business operates profitably, poor eyesight is corrected, the bus arrives on time, or telephones transmit calls rapidly and efficiently. Students deserve a considered, concrete response to their question. One answer is to show them—or better yet, engage them in—mathematics at work.

Endnotes

1. Almost 73 percent of 1992 public high school graduates completed three or more years of mathematics (see Karen Levesque et al., *Vocational Education in the United States: The Early 1990s* (Washington, D.C.: National Center for Education Statistics, U.S. Department of Education, November 1995), Table 40, 117); 54.5 percent of 1992 public high school graduates completed Algebra II, 68.6 percent completed geometry, 19.5 percent completed trigonometry, and 8.8 percent completed calculus (see *The Condition of Education, 1995,* Washington, D.C.: National Center for Education Statistics, Table 26-3, 267).
2. *Curriculum and Evaluation Standards for School Mathematics,* (Reston, Va.: National Council of Teachers of Mathematics, 1989).
3. Samuel Bowles and Herbert Gintis, *Schooling in Capitalist America: Education Reform and the Contradictions of Economic Life* (New York: Basic Books, 1976).
4. This example is drawn from William K. Kershner, *The Advanced Pilot's Flight Manual* (Ames, Iowa: Iowa State University Press, 1984), 4-5.
5. For examples of high schools pursuing some interesting new approaches to work-related curricula, see "Seventh Annual Business Week Awards for Instructional Innovation," *New American High Schools: Preparing Students for College and Careers* (New York: McGraw-Hill, May 1996).
6. For more on using industry themes to organize education around work, see Gary Hoachlander, "Industry-Based Education: A New Approach for the School-to-Work Transition," in *School-to-Work: What Research Has to Say About It* (Washington, D.C.: U.S. Department of Education, Office of Educational Research and Improvement, June 1994).
7. For more case studies of high schools that have developed these strategies, as well as assistance in how to organize curricula and instruction around work, see Mikala L. Rahn et al., *Getting to Work: A Guide for Better Schools* (Berkeley, Calif.: National Center for Research in Vocational Education, University of California, Berkeley, 1996).
8. Mikala L. Rahn, *Profiles of the National Industry Skills Standards Projects* (Berkeley, Calif.: National Center for Research in Vocational Education, University of California, Berkeley, MDS-881, June 1994); Thomas Bailey and Donna Merritt, *Making Sense of Industry-Based Skill Standards* (Berkeley, Calif.: National Center for Research in Vocational Education, University of California, Berkeley, MDS-777, December 1995).
9. *Applied Academic and Workplace Skills for Automobile Technicians* (Herndon, Va.: National Automotive Technicians Education Foundation, 1995).

Mathematical Competencies that Employers Expect

ARNOLD PACKER
Institute for Policy Studies, Johns Hopkins University

Today's mathematics classrooms stress skills that few students need while neglecting skills that employers really want. In this competitive age, mathematics education must provide skills students will use ten years after they graduate.

Customer satisfaction has become the primary goal of America's most successful corporations. Like every other part of the contemporary U.S. economy, education (including mathematics and science) will have to become more customer driven. There is, however, one major difference: unlike the rest of the economy, education's customers are somewhat elusive. Although some suggest that employers are the ultimate customers, Peter Drucker's view may be closer to the mark: "The result of school is a student who has learned something and put it to work ten years later."[1] Higher education, with its eye on alumni contributions, should feel very comfortable with this notion of a "virtual customer"—not the youth sitting in the lecture hall but the imagined successful adult ten years later.

As is true of education's customers, the notion of "success" is also an elusive concept. Success must be defined broadly. Not only should successful adults have the mathematical skills needed to satisfy job requirements but they must also have the competencies needed to fulfill other roles. As good citizens, they should be competent enough to understand the public issues of the day. As lifelong learners, they do not want mathematical deficiencies to stand in the way of other intellectual pursuits. In some idealized sense, their mathematics education should maximize the chances that they will have the tools necessary in the future to solve whatever real problems they will encounter on the job or elsewhere.

In a practical sense, students need to focus on more immediate goals: passing the mathematics courses required for the degrees or other credentials that serve as gatekeepers to their chosen careers. Yet, as we shall see, much of the curriculum in these required courses is irrelevant to most careers. Worse yet, many Americans do not have the mathematical skills they need for work and life. The solution may require that schools abandon a failed curriculum that emphasizes techniques (algebra, geometry, trigonometry, and calculus) but relegates applications to an afterthought. Far better to start with problems that the class of 2000 will face in 2010 and deduce from this goal what techniques students need to master.

What the Job Market Demands

Thus we are led to ask: just what mathematics do employers actually want? Current methodologies for determining the skills needed for any particular job usually include an analysis of frequency and criticality (sometimes called importance). How often does the need for the skill arise and how much is at stake if it is lacking? This methodology can be expressed mathematically. The importance of any skill in a particular job is some function of the probability of its being used, weighed by its criticality in performing the task. Taking this formulation across jobs for any particular student, the market demand D for any skill is a function of the probability P of having any job multiplied by the frequency F of its being used, weighed by its criticality C and the economic importance E of the task. This is, demand D is a function of P, F, C, and E.

Although we do not have the supporting data to estimate this function statistically, mathematical reasoning can guide our search for the mathematics competencies needed in the marketplace. We can then compare the demand for mathematics skills to the mathematics taught in the current curriculum. This comparison will show which competencies are overemphasized and which deserve more attention.

For example, Table 1 shows the proportion of various jobs in the labor force for which mathematics is likely to be important as well as the proportion of some jobs that are not likely to require much mathematics. As this table makes clear, jobs that require a lot of mathematics represent but a small fraction of the labor force. In other words, the probability P is relatively small for jobs that require higher order mathematical skills. Economists, mathematicians, physicists, and astronomers combined account for only some 80,000 jobs. Add 1.5 million engineers, nearly one million accountants, and approximately a half-dozen other jobs high in mathematics content and the total is less than 5 million in a total labor force of about 125 million. Indeed, the total of all 4.74 million jobs characterized by high mathematics content is slightly fewer than the 4.75 million employed in retail sales, only one of the selected low-math content jobs.

Table 1.
Proportion of Selected Jobs in U.S. Labor Force.

Selected Occupations	*Employment* (Thousands)	*Employment* (Percentage)
High Mathematics Content:		
Engineers	1,519	1.3%
Accountants/Auditors	985	0.8%
Financial managers	701	0.6%
Computer programmers	565	0.5%
Computer system analysts	463	0.4%
Clinical laboratory technicians	258	0.2%
Pharmacists	169	0.1%
Economists	37	0.3%
Mathematicians	22	0.2%
Physicists/Astronomers	20	0.2%
Total	4,739	4.0%
Low Mathematics Content:		
Retail sales workers	4,754	4.0%
Waiters/waitresses/bartenders	4,400	3.7%
Secretaries	3,576	3.0%
Chefs/Cooks	3,100	2.6%
Janitors	3,000	2.6%
Office clerks	2,737	2.3%
Truck drivers	2,700	2.3%
Cashiers	2,633	2.2%
Total	26,900	22.9%
Total U.S. Labor Force	117,600	100.0%

Source: Occupational Outlook Handbook. Washington, D.C.: U.S. Department of Labor, 1992.

But what about the mathematics needed for more common jobs? Newspapers abound with stories about the increasing need for mathematics in all jobs. The *New York Times,* reporting recently on the resurgence of hiring in the automobile industry, emphasized the attractiveness of jobs that pay over $30,000 annually with full benefits and that range up to $70,000 with overtime. But the *Times* subhead also stressed the new hiring requirements: "This Hiring Spree Is Rewarding Brains, Not Brawn." "At one point we were hiring hands and arms and legs, and now we are hiring total people—with minds more important than the other," reported Robert Eaton, chairman and chief executive of Chrysler.[2] The employment-testing company, Aon Consulting (formerly HR Strategies), screens automobile industry job candidates for reading, mathematics, spatial relations, manual dexterity, and spatial reasoning. Figure 1 shows an example from the agency's test for "Practical Mathematics." This screening test reveals that despite the rhetoric about minds and brains, the automobile industry is not looking for *higher* mathematics. The

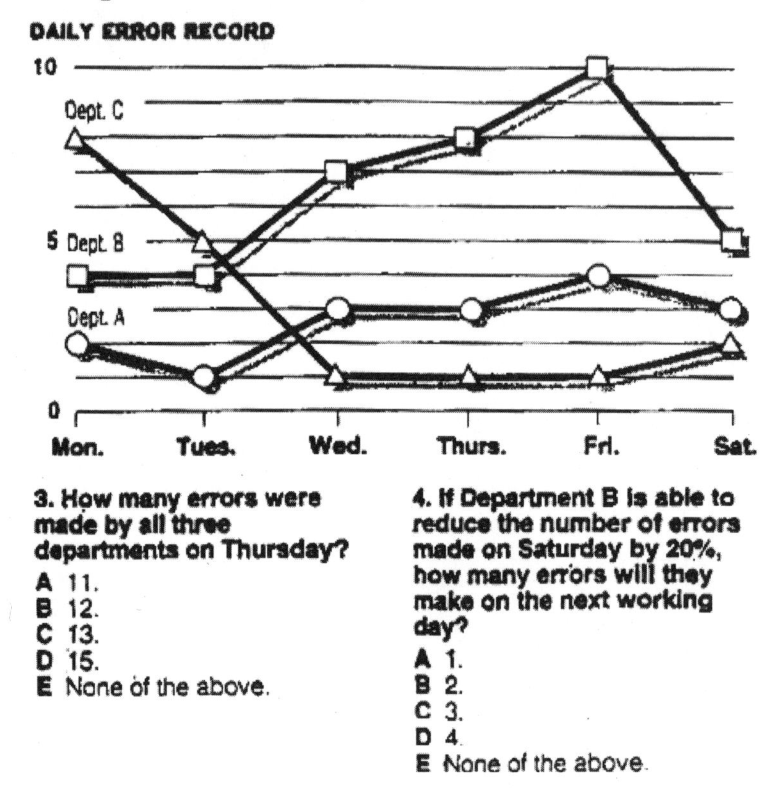

Figure 1.
Typical example from employment screening test.

ability to read graphs, understand percentages, and reason mathematically will suffice.

We might well wonder how representative jobs in the auto industry are of jobs in the U.S. economy as a whole. Is the *Times* story just an anecdote or does it reveal something significant about the nation's mathematics requirements? As with many other things, the answer depends on how the question is asked and who is doing the asking. This issue has been studied through surveys of employers, employees, and job analysts by many organizations, including the Secretary's Commission on Achieving Necessary Skills (SCANS), the American College Testing Corporation (ACT), and the American Institute for Research (AIR). The results, while differing in detail, paint a consistent picture: today's mathematics classrooms stress skills that few students will use while neglecting skills that employers really need.

Data on Required Mathematics

In 1991, SCANS asked employees and supervisors to rank 17 foundation skills and 20 competencies for 35 entry-level jobs (see Tables 2 and 3). Mathematics came in dead last among the foundation skills; arithmetic came in fourteenth. (Of note, no other "academic" course even made the list.) Surely, you say, this relative ranking would increase as the level, pay, and complexity of the jobs increased. If employers were asked instead about the mathematics needed by their engineers or scientists, they would certainly give a different answer. But what about other jobs—the ones that most ordinary people might expect to hold?

Fortunately, ACT is addressing this question more broadly through its National Job Task Analysis, which is determining "generalized work behaviors" (or GWBs) for the entire economy. Its initial sample of 3,000 employees is intended to represent 80 percent of the nation's jobs. Mathematics does not even make the top 25 in ACT's preliminary list of the most important

Table 2.
Foundation Skills Required for Entry-Level Jobs.

Foundation Skills		*Mean*	*Standard Deviation*
F13	Responsibility	4.71	.51
F17	Integrity/Honesty	4.61	.63
F05	Listening	4.61	.64
F06	Speaking	4.33	.86
F01	Reading	4.32	.88
F09	Problem solving	4.20	.92
F15	Social	4.17	.82
F16	Self-management	4.12	.85
F14	Self-esteem	4.11	.82
F11	Knowing how to learn	4.01	.92
F08	Decision making	4.01	1.01
F02	Writing	3.87	1.14
F12	Reasoning	3.87	.94
F03	Arithmetic	3.61	1.18
F07	Creative thinking	3.40	1.12
F10	Seeing things in the mind's eye	3.26	1.24
F04	Mathematics	2.75	1.26

Source: Adapted from "Skills and Tasks for Jobs," SCANS Report for America 2000, U.S. Department of Labor, 1992.

Table 3.
Competencies Required for Entry-Level Jobs.

Competencies		*Mean*	*Standard Deviation*
C09	Participates as a member of a team	4.24	.94
C01	Allocates time	4.19	.96
C11	Serves clients/customers	4.15	1.34
C07	Interprets and communicates information	4.00	1.15
C15	Understands systems	3.77	1.10
C14	Works with cultural diversity	3.72	1.21
C10	Teaches others	3.67	1.19
C12	Exercises leadership	3.63	1.18
C06	Organizes and maintains information	3.61	1.25
C05	Acquires and evaluates information	3.59	1.24
C13	Negotiates to arrive at a decision	3.39	1.21
C16	Monitors and corrects performance	3.35	1.19
C19	Applies technology to task	3.35	1.46
C17	Improves and designs systems	3.11	1.31
C08	Uses computers to process information	3.00	1.70
C18	Selects technology	2.89	1.45
C03	Allocates material and facility resources	2.82	1.32
C04	Allocates human resources	2.82	1.48
C20	Maintains and troubleshoots technology	2.70	1.45
C02	Allocates money	2.21	1.37

Source: Adapted from "Skills and Tasks for Jobs," SCANS Report for America 2000, Washington, D.C.: U.S. Department of Labor, 1992

GWBs. Even arithmetic is only in slot number 14. A list of the top 25 GWBs is shown in Table 4.

An earlier ACT analysis specifically examined the need for applied mathematics. The ACT sample of 583 jobs included technical jobs (65 percent of the sample) and careers in science (7 percent of the sample, most of them in engineering and related technologies). ACT researchers divided mathematical skills into seven levels. More than 90 percent of the jobs profiled could be performed with no more mathematics than required at level 6: "Using negative numbers, ratios, percentages, and mixed numbers . . . calculate multiple rates, find areas of rectangles and circles, volumes of rectangular solids, solve problems involving production rates and pricing schemes." Reaching level 6 required only enough algebra to look up formulas or transpose a formula before calculating. Only 18 percent of the jobs profiled were actually at level 6; nearly three-quarters of all jobs were not even at this level.

About 8 percent of the jobs required level 7, but even this bar was not very high. Level 7 requires the ability to calculate the volumes of spheres, cylinders, and cones, solve nonlinear equations, apply basic statistical con-

Table 4.
25 Top Generalized Work Behaviors Arranged by SCANS Skills.

A. Planning and allocating resources such as scheduling, budgeting, and allocating space and staff.

4. Schedule work activities for oneself.
6. Determine the priority of work activities.
15. List things that need to be done or remembered for work.

B. Analyzing, evaluating, organizing, and communicating information, including using computers for these purposes.

2. Use a computer to locate, process, or communicate information.
5. Provide information to people who ask specific questions or have specific concerns.
8. Prepare paperwork necessary for a work activity.
16. Check documents for completeness and quality of appearance.
17. Judge the importance, quality, and accuracy of information.
20. Enter information or data into computer files.
21. Read forms, work orders, requests, or instructions to determine what work needs to be done.
23. Use hand signals or short vocal instructions to control a work operation.

C. Interpersonal skills: working on teams, negotiating, leading, teaching, serving customers.

1. Listen to the concerns of clients/customers and respond.
11. Work with people in other departments to complete tasks.
13. Keep order/discipline among groups of people.
18. Take customer orders for equipment/machinery, materials, or services.
22. Coordinate own work activities with the activities of others.
25. Listen to instructions from or concern of supervisors or co-workers and respond.

D. Using and making decisions regarding technology.

9. Operate individual pieces of equipment/machinery.
24. Operate computerized equipment other than a personal computer.

E. Understanding, monitoring, and improving systems.

3. Safeguard information, money, valuables, buildings/property, or people.
7. Manage the activities or resources of a department, business, organization, or facility.
10. Monitor the health condition of people.
12. Monitor the activities or attitudes of clients/customers to make sure they are satisfied.
19. Monitor work processes or operations to make sure there are no problems.

Basic skills—arithmetic and mathematics

14. Perform arithmetic (addition, subtraction, multiplication, and division) as part of work activities.

Note: Italic headings refer to SCANS competencies; numbered work behaviors indicate the relative importance of each GWB as determined by ACT.

cepts, and locate errors in multiple-step calculations. Calculus, vectors, and even serious trigonometry were not mentioned despite their importance in today's mathematics curriculum.

In addition to ACT, relevant information is available from the Department of Labor's O*NET project and its contractor (AIR), which is developing a

Table 5.
Anchors for AIR Levels of Mathematical Skill and Knowledge.

Skill:	
Level 6:	Develop a mathematical model to simulate and resolve an engineering problem.
Level 4:	Calculate the square footage of a new home under construction.
Level 2:	Count the amount of change for a customer.
Knowledge:	
Level 6:	Derive a complex equation (requires advanced concepts such as calculus, nonlinear algebra, and statistics).
Level 4:	Analyze data to determine the geographic area with the highest sales.
Level 1:	Add two numbers.

data base of jobs in the U.S. economy. During the development of this data base, the O*NET project surveyed employees (job incumbents) representing about 35 percent of U.S. jobs and occupational analysts examined over 1,000 jobs representing the universe of U.S. occupations. For each job, both incumbents and occupational analysts were asked:

- What level of mathematical *skill* is needed? (Skill is defined as the ability to use mathematics to solve problems.)
- How much mathematical *knowledge* is needed? (Knowledge refers to techniques such as arithmetic, geometry, algebra, calculus, and statistics.)

AIR also designed a seven-point scale to measure the levels of skill and knowledge required for an occupation. Three anchors were provided for each item (examples are shown in Table 5) and raters were asked to perform their ratings in relation to these anchors. The occupational analysts found that only 13 of the 1,122 occupations examined required high levels of mathematical skill and knowledge (above 5.0 on the seven-point scale). In comparison, 452 occupations (approximately 40 percent) did not require either mathematical skill or knowledge above the low range (the 2.5 level), while 356 (32 percent) required both skill and knowledge in the mid range (see Table 6). The remaining occupations required mid-range levels of either mathematical skill or knowledge, but not both (Table 7).

The SCANS, ACT, and AIR analyses provide consistent answers to the question of what mathematics is needed on the job. Insofar as their careers are concerned, over 90 percent of students would hardly ever be disadvantaged if they never went further than fairly simple geometry and a modest course in algebra. A small amount of solid geometry, somewhat more advanced algebra, and some statistics would suffice for most of the remain-

Table 6.
Classification of Jobs by Mathematical Skill and Knowledge.

		Mathematical Skill		
	Mean	*0.0 - 2.5*	*2.5 - 5.0*	*5.0 - 7.0*
Mathematical knowledge	0.0 - 2.5	452	255	0
	2.5 - 5.0	33	356	10
	5.0 - 7.0	0	3	13

ing 10 percent. Taking integrals or solving differential equations is needed primarily by those who teach college mathematics and for a smattering of advanced engineering or scientific jobs. Today's extraordinarily high unemployment rate for mathematicians with Ph.D.'s (as much as 14 percent by some estimates) shows clearly that the supply exceeds the demand for people with advanced mathematical education. Yet in 1989 the National Council of Teachers of Mathematics (NCTM) proposed standards for grades 9 to 12 that include such "basics" for all students as "trapezoidal estimates of the area under a curve"![3]

Table 7.
Mathematical Skill and Knowledge Needed in Selected Jobs.

As rated by employees (job incumbents):

Low skill, low knowledge: Police patrol officers; Cooks; Food preparation workers; Nursing aides, Orderlies and attendants; Janitors and cleaners; School bus drivers.
Medium skill, low knowledge: Stock clerks, Sales floor workers; Teachers' aides and assistants; Clerical workers.

Low skill, medium knowledge: Tellers; Receptionists and information clerks; Waiters and waitresses.

Medium skill, medium knowledge: Education administrators; General managers; Top executives; Computer programmers; Elementary school teachers; Registered nurses; Medical clinical laboratory technologists; Salespersons; Cashiers; First-Line supervisors; Secretaries; Bookkeeping, accounting, and auditing clerks; Police and Detective supervisors; Maintenance repairers; Packaging and filling machine operators; Tractor trailer truck drivers.

High skill, high knowledge: Earth drillers.

As rated by job analysts:

Low skill, low knowledge: Composers; Musicians; Dancers; Fire fighters; Police investigators; Bakers; Child care workers; Farm workers; Taxi drivers and chauffeurs.

Medium skill, medium knowledge: Administrative managers; College and university administrators, Educational program directors; Government service executives; Accountants; Biologists; Teachers; Social workers; Farmers; Commercial airplane pilots; Civil engineers; Construction and building inspectors; Doctors of medicine; Police and detective supervisors; Loan officers and counselors.

High skill, high knowledge: Aeronautical, nuclear, and computer engineers; Physicists; Statisticians.

Mathematics Competencies in Short Supply

It would be wrong to conclude from these findings that most U.S. workers have greater mathematics competence than they need. Although most jobs do not require higher mathematics (more than basic algebra), many mathematical skills (as well as writing and other skills) are in short supply relative to the demands of today's job market. Only 9 percent of an ACT sample of 125,000 examinees tested at level 6 skill or knowledge, yet ACT's job profiles showed that 18 percent of jobs require level 6 performance. Only 2 percent of the examinees tested at level 7, yet 8 percent of jobs require level 7.[4] Consequently, there is an undersupply of mathematically competent employees of 9 percent at level 6 and 6 percent at level 7. According to ACT, about half of all jobs (47 percent) require level 5 or higher, but only about a third (34 percent) of the examinees were that competent, a gap of 13 percent. And what is level 5? Just the ability to determine what information and calculations are needed to convert within systems of measurement (e.g., converting from ounces to pounds) or between systems (from centimeters to inches) or to calculate with mixed units (hours and minutes).

Some years ago, a similar analysis using U.S. Department of Labor (DOL) and Educational Testing Service data produced comparable results.[5] The biggest gap between supply and demand was at the Department of Labor's level 4: "Perform arithmetic, algebraic, and geometric procedures in standard, practical applications, such as shop mathematics and accounting." Jobs requiring more advanced mathematical and statistical techniques accounted for fewer than 10 percent of the total.

In addition to these broad, industrywide studies, the U.S. Departments of Labor and Education recently supported the development of voluntary skills standards for employees in 22 industries. Recent legislation has established a National Skills Standards Board to bring these efforts together in a national system, which yielded much the same picture. High-volume industries such as retailing, hospitality, and health care have modest mathematical requirements that primarily emphasize basic arithmetic. Some of the more technical jobs, such as electronic technician, advanced manufacturing, and computer-aided drafting and design, require basic algebra, geometry, and trigonometry. Yet even these demands are modest. The standards for trigonometry, for example, require only the ability to use a calculator to compute cosine, sine, and tangent. None of the more esoteric trigonometric functions (or any other functions for that matter) or any calculus are even mentioned.

These findings on what mathematics competencies are in short supply are consistent with data on wage rates. A recent study focused on the issue of the growing importance of basic mathematics by asking whether mastery of basic mathematics by people graduating from high school in 1980 played a larger role in determining their wages at age 24 than it did for people graduating in 1972. This study found that the skills needed were the "ability to follow directions, manipulate fractions and decimals, and interpret line graphs." The authors conclude that "a high school senior's mastery of skills taught in American schools *no later than the eighth grade* is an increasingly important determinant of subsequent wages." (Emphasis added.)[6] Indeed, the premium placed on good basic mathematics skills for high school graduates who did not go to college roughly doubled over the eight years (from 22 cents to 53 cents per hour for males and from 39 cents to 74 cents for women).

Despite this evidence, the overriding goal of the mathematics departments of many secondary schools and community colleges seems to be to train students to understand the calculus of trigonometric functions. Higher education, especially engineering schools, represents the "customer" for these mathematics teachers—a vestige of an earlier era when engineering and surveying jobs were the likely outcomes for the mathematically talented. Today, this approach does not reflect reality. Although some teachers may talk about studying calculus for its concepts (e.g., to understand ideas such as rates of change), these concepts are rarely on the test. Good grades are more likely to depend on remembering clever transformations that a few students may use in engineering studies but that most students forget soon after the course is over.

This emphasis on trigonometry and calculus comes at the expense of solid grounding in applying more common mathematical techniques. It leaves no room for the new mathematics requirements of the changing work place (such as statistical quality control or computers) that now extend well beyond engineering. (There are even questions about the importance, for most engineering careers, of calculus, which is often required at the expense of learning more relevant mathematics. See, for example, Ferguson's telling criticism that engineering education emphasizes esoteric analysis at the expense of expertise in design and solving practical problems.)[7]

To summarize, the implicit goal of today's mathematics curriculum is unnecessary for the vast majority of U.S. workers. At the same time, the mathematical preparation of most graduates is inadequate. Mathematics educators need a different approach to the questions of what mathematics to teach and how to teach it. Clearly, many leaders in the field recognize this need and there are ongoing efforts to find a remedy. So far, however, the response has been inadequate.

A Problem-Centered Curriculum

Clearly, this analysis is rather discouraging for those who, like the author, believe in the value of rigorous mathematics education. What if the question was not "What mathematics *is* being used on the job?" but instead, "What mathematics *should* be used and how?" The mathematician Henry Pollak has thought deeply about this question based on years of experience in one of America's leading industries—telecommunications. Pollak's list of employer expectations[8] is a good place to start thinking about a new approach to the mathematics curriculum. High-performance employers want to hire men and women who:

- Are able to set up problems;
- Know a variety of techniques that apply to problems;
- Understand the mathematical features of problems;
- Work with others on problems;
- See how to apply mathematical ideas to problems;
- Are prepared for open, unstructured problems; and
- Believe in the use and value of mathematics in problem solving.

Pollak's formulation leads to the conclusion that employers are less concerned with the mathematics their employees can do (AIR's knowledge) than with the problems they can solve (AIR's skills). Employees prepared in this manner can help high-performance firms throughout the nation that are restructuring to become more competitive. Irrespective of the buzzwords used, the changing nature of the work place means more analysis, decision making, and problem solving by so-called "front-line" workers. The relevant questions are: What is being analyzed? What decisions are being made? What problems are being solved?

The SCANS Commission concluded that graduates who can solve certain general problems within the specific domains of their industry will be well prepared for the twenty-first century. It identified five categories of such problems:

1. *Planning Decisions.* Planning and resource allocation decisions such as scheduling, and budgeting and allocating space and staff.
2. *Information.* Analyzing, evaluating, organizing, and communicating information, including using computers for these purposes.
3. *People.* Interpersonal skills, such as working on teams, negotiating, teaching, and customer service.
4. *Technology.* Using and making decisions regarding technology.
5. *Systems.* Understanding, monitoring, and improving systems.

These categories suggest that an effective mathematics curriculum would meet the following criteria:

- Courses should meet the general goals outlined in the NCTM *Standards:* students learn to value mathematics, to become confident in their own ability, to become mathematical problem solvers, to communicate mathematically, and to reason mathematically.[9]
- All students should be required to take enough mathematics to qualify for a large share of jobs and to be prepared to go further in their studies. For example, if all high school students took two years of mathematics—algebra I and geometry—they would reach a level sufficient to qualify for 75 percent of the available jobs. If they chose to take three years of mathematics, they should have the mathematics needed for 90 percent of the jobs in the U.S. economy.
- Mathematically inclined students should be able to go as far as they want in fields that use mathematics. Note that problems in four of the five SCANS areas (planning decisions, information, technology, and systems) are represented in the mathematics curriculum at levels as high as the Ph.D. So students should not have to choose between academic and vocational tracks. All students should be required to study in a single track until they meet the minimum requirements, then continue if they elect to do so.
- Mathematics courses should be organized not around techniques but around identifying and solving problems in the five SCANS competency domains. In other words, instead of being organized around mathematical themes such as factoring polynomials or proving theorems, algebra and geometry should be organized around planning, information, and systems problems.

Choosing Relevant Problems

In a sense, all of today's mathematics courses teach students to solve problems: simple addition, long division, taking square roots, factoring polynomials, proving triangles congruent, and so on. Yet despite a plethora of "problems," today's curriculum does not address the realistic, everyday problems that students will face as adults. Here, for example, are five (not quite) randomly selected problems listed in the NCTM *Standards*[10]:

- Find five examples of numbers that have exactly three factors (p. 93);

- Approximate the area of the region under the curve $y = 2x$ above the x axis and between the lines $x = 1$ and $x = 3$ (p. 148);
- Find the roots of the equation $5x^3 - 12x^2 + 8 = 0$ (p. 152);
- Investigate limiting processes by examining infinite sequences (p. 180); and
- Suppose a Ferris wheel with a radius of 25 feet makes a complete revolution in 12 seconds. Develop a mathematical model relating the height of a rider to time (p. 184).

None of these problems—which, by the way, NCTM suggests are to be solved by *all* students—seem to be related to the requirements described by SCANS, ACT, AIR, DoL, or in the new industry skill standards. NCTM's suggested problems are not at all related to employer requirements. Although the Ferris wheel problem, unlike the other four, may seem to relate to the "real world," very few workers need to solve such problems in today's economy. The problems students are expected to solve should not only relate to the real world but should be something people would realistically encounter frequently on the job or in their roles as citizens or parents.

Imagine, instead, alternative strategies for teaching mathematical problem solving based on the following criteria:

- Mathematical techniques are identified in the process of searching for solutions to work-related problems. This reverses the usual situation wherein "applications" are an afterthought designed to illustrate a mathematical technique, inserted at the end of the chapter, and clearly artificially constructed.
- Problems often take some time to solve and can be solved in groups, and the results can be presented to the class. This is in contrast to the type of mathematical problems that can be solved in 10 to 20 minutes in a mathematics test taken by solitary students. (Cooperation, in the traditional test environment, is viewed as cheating.)
- Solutions call on mathematics in an authentic way, yet the mathematics is rigorous enough to satisfy college entrance requirements.
- Work-related problems should be important to a wide variety of jobs. At a minimum, the student might expect to have a one-in-a-hundred chance of running into a problem of this type (that is, it should be a problem that is routinely solved by over one million workers in the U.S. economy).

Think of what would happen to U.S. mathematics if this 1 percent criterion were applied. Every assigned problem would have to represent a class of problems being solved by 1.25 million current workers!

SCANS Competencies and Generalized Work Behaviors

Table 4 associates SCANS competencies with ACT's 25 most common GWBs. At first glance, it is hard to see where mathematical or quantitative issues fit into this schema. But quantitative literacy skills are hidden beneath the surface of many of these job requirements, and can easily be elicited by creative teachers and innovative texts. For example, here are some pairs (SCANS competency first, GWBs second) with suggested relationships to mathematical and quantitative literacy:

- **Planning and allocating resources:** *Scheduling oneself and setting priorities of work activities.* In some cases, students would require only simple arithmetic to develop schedules or set priorities, but in other situations the solution could demand algebra to create visual representations (e.g., Gantt diagrams or PERT charts) for activities whose completion times are uncertain.
- **Use information:** *Provide information to people; use a computer to process or communicate information.* A student may need enough algebra to build formulas into a spreadsheet so that a detailed budget or invoice can be prepared for a dinner for 12 or a construction job.
- **Interpersonal skills:** *Listen and respond to the concerns of clients and customers (or the instructions of supervisors and co-workers); take customer orders for equipment, materials, or services.* Students might have to answer questions such as "What would it cost to have a birthday party for 12 at this restaurant?" Or respond to instructions such as "Give me enough paint to cover the kitchen with two coats."
- **Technology:** *Operate individual pieces of equipment or machinery, including computerized equipment.* Students should be able to read gauges and determine when a piece of equipment needs an oil change or other maintenance. They could be required to compare operating characteristics to information provided in tabular or graphical form.
- **Systems:** *Manage the activities or resources of an organization; monitor work processes and attitudes of clients or customers; safeguard information, money, valuables, buildings, property, or people.* Carrying out these functions may require knowledge of the statistics needed to design a survey or to apply the concepts of statistical quality control. Students should be able to develop and present quantitative data in

tabular and graphical form with enough surrounding prose to make the story intelligible. Developing a model of a process or an activity may also be required.

Case Studies Focusing on Generic Problems

To make all this less abstract, let us consider a set of real-world problems from industry. Representatives of Boeing, the Alliance (between AT&T and the unions that represent its workers), Ford, and the National Association of Manufacturers helped identify some SCANS-type problems that they would expect an associate degree holder to be able to solve.[11] For example, a graduate of a community college competent in allocating money should be able to handle budgets and cost accounting. Competence in technology includes the ability to recommend alternative technologies to meet some need.

These competencies are embedded in the generic problem of allocating financial resources to maximize the profit for a product that contains a new technology, for example, an electric car.[12] Working in groups, career professionals have to budget for research and development, labor, materials, and new production equipment. They have to set the selling price for the product and develop a strategy to optimize the advertising budget. Employees have to read charts, interpret marketing data, express relationships in linear equations and use them in a spreadsheet, and understand how to use inequalities to maximize profit. They will report back (communicate mathematical facts or concepts), and engage in a group decision-making process leading to the selection of a production and marketing plan.

A case study setting up the electric car problem could be used in a number of mathematics courses as well as in interdisciplinary courses. Similar real work-world case studies can help a student learn how to deal with systems while introducing the concepts of statistical process control and the statistics needed to use this technique. Another case study, also dealing with systems, asks students to examine how the broader environmental and social system constrains plant locations and design decisions. The mathematics can get quite advanced because students are required to work with equations that describe the physics of plumes emitting from smokestacks. As the electric car offers one example of a generic problem (introducing a new product), the smokestack represents a class of plant location problems focusing on environmental concerns.

Case studies that set up complex real-world situations as the focus for the mathematics curricula can be used in high schools as well as in community colleges. For example, students may be asked to do a budget analysis for a new physical therapy service and develop a schedule for bringing it on line, thereby learning how to plan and allocate resources. Or they may be asked to design, budget, and schedule the construction of a retail store, applying mathematics to the retail and built-environment fields. In the latter case, they may have to understand and design a system of traffic flow to yield a target rate of revenue. Students will have to communicate their mathematical findings (an information competency) when presenting their results. Or they may be asked to develop a budget for a tour to South Africa. In all case study problems, students work together, thereby learning teamwork skills as well as valuable mathematical problem-solving skills.[13]

Teaching Mathematics with an Eye on the Work Place

The point of all this is to have students set up, solve, and present solutions to open-ended problems. The case study approach meets the goals set out by NCTM (although it would replace the NCTM examples of typical problems with those described above). It reconciles the current practice of U.S. (but not European) mathematics education of labeling courses with names that refer to broadly defined mathematical techniques—algebra, geometry, and trigonometry—with the emphasis employers place on work-based competencies.

Curriculum developers may ponder whether learning *area* = *length* × *width* is an algebra or a geometry problem. Employers would rather know whether their workers can allocate space in an efficient manner. Of course, solving the allocation problem may very well require knowing the equation for area. By using the language of the work place rather than that of the school, educators may be more successful in engaging businesses as partners in the education process. At a minimum, it may increase the attention that employers pay to school transcripts.

More important, by stressing applications educators can reduce the gap between the current mathematics competency of most workers and that needed for twenty-first-century problem solving. Mathematics courses should emphasize the skills needed to solve problems on the job. One possible standard might be that two years of high school mathematics should equip students for 75 percent of the jobs and three years for 90 percent. This criterion needs more study, but defining mathematics standards in these terms is sensible. In this new competitive age, mathematics education must provide students with what the job market demands—the ability to solve common and important problems.

The ACT and AIR research is not yet complete and the nation is just beginning to develop industry skill standards. Although the SCANS Commission likes to think that most of its competency scheme will endure—especially the broad requirements for planning, using information, interpersonal skills, technology, and systems—surprises are inevitable. Irrespective of the pace or extent of change, however, there are certain concepts that should shape mathematics education.

Mathematics education should develop competencies to solve problems that are frequently found in the work place, not train 100 percent of our students for the minuscule percentage of jobs held by professional users of mathematics. Yes, there are uses for mathematics other than in the work place. Mathematical aesthetics is one, but on this terrain mathematics needs to compete—as an elective—with fine arts, philosophy, and music. Good citizenship also requires mathematics, but statistics easily outpaces trigonometry or calculus as the mathematics needed to understand public policies. Simply put, the mathematics curriculum most in evidence currently is not optimally aligned with the needs of the class of 2000 and beyond. A better match will be achieved when educators start paying more attention to their customers—students ten years after graduation.

Endnotes

1. Peter Drucker, *The New Realities* (New York: Harper and Row, 1989).
2. Robyn Meredith, "New Blood for the Big Three's Plants," *The New York Times* (Sunday, April 21, 1996, Section 3: Financial Desk, 1).
3. *Curriculum and Evaluation Standards for School Mathematics* (Reston, Va.: National Council of Teachers of Mathematics, March 1989, 148).
4. "Job Skill Comparison Charts," (Iowa City: ACT, January 1996, 7).
5. Arnold H. Packer, *Preparing for Work in the Next Century* (Indianapolis: Hudson Institute, May 1990).
6. Murnane et al, "The Growing Importance of Cognitive Skills in Wage Determination," (NBER Working Paper No. 5076, *NBER Digest,* October 1995).
7. E. S. Ferguson, *Engineering and the Mind's Eye* (Cambridge, Mass.: Massachusetts Institute of Technology Press, 1992).
8. *Curriculum and Evaluation Standards for School Mathematics* (Reston, Va.: National Council of Teachers of Mathematics, March 1989, 4).
9. Ibid.
10. Ibid.
11. Arnold H. Packer, "An Associate Degree in High Performance Manufacturing," Final Report to the Sloan Foundation, November 1994 Johns Hopkins University.
12. Under a grant from the National Science Foundation, the author is creating a CD-ROM-based "electronic case study" for a problem such as the one described. The module is designed to be part (about 20 percent) of a college algebra course.
13. The author is working with the Baltimore City Public Schools to develop CD-ROM-based courses along these lines under a five-year Technology Challenge Grant from the U.S. Department of Education.

Quantitative Literacy Across the Curriculum

GLENDA PRICE
Spelman College

Quantitative reasoning is a defining skill for success in our increasingly technological world. Although courses that emphasize applications in real-life situations can help, a better model might be "mathematics across the curriculum."

As we approach the twenty-first century, computers and technology have made our world increasingly analytical. Both everyday decisions and challenges in the work place demand greater quantitative skills than in the past. Today, no one can be considered educated who is unable to interpret data, perform fundamental computations, make comparisons, and assess risks.

Some analysts suggest that quantitative literacy is one of the defining skills that determine who will be successful and who will not. Solutions to global problems such as the environment, population growth, and the economy all depend on quantitative skills. For the women of Spelman College, the overwhelming majority of whom are African American, quantitative literacy may make the difference between advancing in today's world or remaining trapped in low-paying positions with little prospects for advancement.

Spelman College educates future African-American female leaders by providing a broad liberal arts foundation with sufficient depth for success in today's world and sufficient breadth for growth in the twenty-first century. Every Spelman student is provided opportunities and encouragement to gain an appreciation of the arts, an introduction to a second language, an understanding of African Americans in history, a grounding in the social sciences, and—most important for this discussion—scientific, quantitative, and computer literacy. At Spelman, quantitative literacy means understanding the mathematical concepts and skills that are necessary for everyday life.

Quantitative literacy is a concern of most faculty, but especially of those in the social sciences for whom quantitative skills are necessary to grasp the concepts of their disciplines. In addition, staff in the Office of Student

Affairs frequently notice the lack of quantitative literacy as a contributing factor in difficulties that students encounter. These range from making bad decisions about credit cards, to difficulties in developing budgets for student organizations, to the inability to estimate the time needed to travel from one place to another. These experiences lead to calls for "practical" mathematics for all students. All faculty and staff, whatever their academic division or administrative perspective, believe that it is the responsibility of the mathematics faculty to develop students' quantitative skills.

Contemporary Mathematics

In many ways, Spelman College is a unique institution. Although Spelman offers a traditional liberal arts curriculum, nearly four of every ten students major in science (chemistry, computer science, biology, mathematics, physics, or engineering). These students approach mathematics as a foundation tool necessary for success in their majors and in their future careers. They generally enroll in several advanced mathematics courses relevant to their majors and thus quickly become quantitatively literate.

Most other students enroll in Math 107: Contemporary Mathematics, a course designed to move students with majors outside the sciences toward quantitative literacy. Contemporary Mathematics emphasizes applications and problem solving in real-life situations, with a particular focus on modeling in management and the social sciences. The course provides, according to the prospectus, "an explicit answer to the question, 'What good is mathematics?'" It presupposes that students had only a year of basic algebra, requires students to use calculators, and introduces such topics as consumer mathematics, linear functions, probability, statistics, and linear programming. After completing this course, a student should understand the mathematics of basic finance such as interest on credit cards and home mortgages; descriptive statistics including mean, median, variance, standard deviation, and the normal distribution; basic probability; and the concept of optimization as illustrated by linear programming.

In the context of a four-year liberal arts education, Contemporary Mathematics provides a good introduction to quantitative literacy—but it is only an introduction. The number of other courses that students have to take for graduation limits our effort in this area to a one-semester endeavor. Yet to become part of a student's intellectual repertoire, the concepts and skills introduced in this course must be reinforced by repeated use in other courses. Thus our faculty encourages students to complete Contemporary Mathematics during their first two years so they can build on their knowledge and skills in subsequent courses in other departments.

Student Experiences

Spelman students are in most respects typical of today's college students. There are, however, some real differences that are undoubtedly a function of their backgrounds and experiences. The 1,900 young African-American women at Spelman are competitive with the best students at institutions across the country. Most Spelman students are highly goal oriented and determined to succeed, and hold high expectations for the future. As a result, many are quite risk averse: they avoid courses and activities in which they believe they have the potential for failure. Thus some students in majors that do not require mathematics as a prerequisite simply delay until the last possible moment completing those graduation requirements that they fear. These students miss valuable opportunities to reinforce what is already a minimum level of quantitative literacy.

Students of many different mathematical backgrounds take Contemporary Mathematics. Those with strong backgrounds have no problem with the course, but those who in the past have experienced difficulty with mathematics will usually require (and seek) special help from their instructors. There are many reasons why students may have trouble in a college mathematics course, among them attitudes about mathematics that are shaped well before students enter college. Although many groups, for example the National Council of Teachers of Mathematics (NCTM), are seeking to improve school mathematics, the effectiveness of these changes will not be seen for many years.

Ironically, even as some students fear Math 107 for its perceived difficulty, others see it as insubstantial. Spelman's achievement-oriented milieu encourages a perception on the part of some students that this course is not sufficiently challenging. They believe that the course is beneath them, that it is a course for students who cannot do "real mathematics," that its very title—Contemporary Mathematics—connotes an inability to grasp the essential elements of mathematics. Indeed, many Spelman students can certainly master a higher-level course, yet most find that Contemporary Mathematics provides just the right combination of motivation and challenge.

Many faculty, especially in the social sciences, also want more for their students. They want students to be able to arrive at logical conclusions even in the face of conflicting evidence; to interpret graphs and charts that they encounter on a daily basis; to determine for themselves the validity of claims without relying on someone else's judgment; to assess claims and detect fallacies; and to express causal relationships. The ability to understand such concepts as population growth, national debt, risk management, and statistical sampling is very important for decision making and problem solving within the social and managerial sciences.

Mathematics Across the Curriculum

Clearly, it takes more than one course to make students quantitatively literate. Other courses and graduation requirements contribute a great deal to the process of developing students' quantitative literacy. The parallel requirement to demonstrate computer literacy is just as important. The logic of computers is an essential element of quantitative reasoning, and all students must be able to use the computer as a tool for computation, organization, analysis, and presentation. Another course that makes a significant difference in the way students think is Interdisciplinary Science—a topically oriented course that is inquiry based, lab rich, and hands on. Students in this course are presented with problems, data sets, and theories; their task is to identify solutions and reach conclusions. Only students who are quantitatively literate can deal adequately with the challenges presented by this course.

Just as the development of good writing skills requires practice in many areas, so quantitative literacy requires stimulation in more than just one or two courses. It is natural, therefore, to encourage "mathematics across the curriculum" in the spirit of the many successful "writing across the curriculum" programs that focus on writing in most courses in each major. Similarly, courses within each major that emphasize quantitative methods would greatly enhance the kinds of experiences many students need in order to develop quantitative literacy.

Of course, faculty in the natural sciences have already integrated quantitative techniques into many of their courses. But what about the arts and humanities? Opportunities certainly exist to include geometry in art classes, inference in psychology, and probability in management. Although faculty in these disciplines frequently use and teach these mathematical concepts, they rarely emphasize them as aspects of quantitative reasoning.

The concept of mathematics across the curriculum is intriguing for many faculty. However, as with writing across the curriculum, many faculty doubt their preparedness to be effective instructors. Most social science faculty are more comfortable than humanities and fine arts faculty in their belief that they can implement instructional programs that foster quantitative literacy.

As the college now holds workshops to assist faculty in designing and evaluating writing exercises for their courses, so a similar model could help faculty in all disciplines provide a special emphasis on quantitative skills appropriate to their disciplines. This kind of program, which embeds mathematical themes in courses across the curriculum, would introduce a multidimensional approach to quantitative literacy that extends well beyond what can be accomplished in Math 107 or in any other single course.

Although faculty from many other departments need to provide support and reinforcement, mathematics faculty are the key to students' quantitative literacy. They must be deeply involved in all aspects of the identification and cultivation of quantitative literacy. They must assist their faculty peers as well as their students. And they must communicate both in word and deed that quantitative reasoning and decision making in everyday life are as important as theoretical mathematics.

Faculty Commitment

The challenge of quantitative literacy at a college such as Spelman is not unlike the challenge faced by every educational institution: to provide offerings of sufficient depth to engage students of diverse interests, and to make such offerings appealing even to students who, often for good reason, are afraid of anything that seems mathematical. Here are some strategies that Spelman faculty have found to be helpful:

- Give faculty freedom to innovate. Encourage creativity. Allow faculty to develop new courses about which they are excited and which therefore will excite their students.
- Link courses to other disciplines. Use examples provided by other fields. Use the media and current events to document the real-world relevance of courses.
- Make classes interactive. Group projects create an environment in which strong students assist weaker ones. Healthy competition between groups pushes everyone to do their best—in presentation as well as in content.
- Create opportunities for student success. Whether working in groups or independently, students should always be able to gain a feeling of accomplishment. Once a student has achieved a certain level of success, confidence builds and more success follows.
- Overcome resistance to change. Many students resist new things. Group projects that are familiar yet contain something not done before can help overcome this natural resistance.
- Take nothing for granted. Everyday examples may not be meaningful to all students. One Spelman student, for example, was mystified by a class on probability that relied on the structure of a deck of cards, since she had never had any experience with cards.
- Show mathematics in action. Try to recreate in courses the visibility and enthusiasm that radiate from student fairs and projects at events

such as Science Days. Encourage mathematics students to present papers and sponsor poster sessions.
- Ensure a critical mass. Larger groups of students create their own dynamic energy. By bringing small classes together, students can create their own support networks.
- Prepare teachers to teach. College faculty have a responsibility to prepare preservice teachers as well as to sponsor in-service programs to maintain teachers' competence. Effective work with current and prospective teachers will yield stronger quantitative literacy skills among first-year students.
- Encourage faculty to be student friendly. Frequently students see people with Ph.D.'s in mathematics as larger than life—as having special skills, a rarefied few. It is important for students (and other faculty) to see mathematicians as just like themselves.

The mathematics faculty at Spelman truly believe that every student can learn mathematics. Their goal is to reduce students' anxiety and make quantitative reasoning accessible to all, not to filter students for selected majors. This commitment to success is part of the Spelman ethos: all faculty believe that all students can succeed, and all stress that success is guaranteed if students apply themselves.

In comparison with their parents' generation, today's college graduates are expected to have larger vocabularies, broader knowledge of history, greater analytic ability, exposure to a wider range of reading, *and* significantly greater capability to reason quantitatively. Computation, interpretation, inquiry, and application of mathematical concepts are critical to life in the contemporary world. Students need multiple opportunities—and multiple inducements—to build their quantitative literacy skills to a level sufficient for the challenge of the twenty-first century.

Yet, in my opinion, there is not yet clear consensus at Spelman about how to implement the goal of quantitative literacy for all students. We agree that quantitative literacy is important and that it cannot be accomplished in any single course. Like our colleagues at other institutions, we struggle continually to find the right level and approach that will meet the needs of all students. This is not a problem unique to Spelman, because colleges across the country are having trouble devising a suitable mathematics course for nonscience majors. What counts is that our faculty are truly committed to the goal of quantitative literacy for all students, and that all share in the responsibility for achieving this goal.

Afterword: Through Mathematicians' Eyes

SUSAN L. FORMAN
Bronx Community College

To mathematicians, quantitative literacy is not the same as mathematical literacy. It is part of all subjects, the responsibility of teachers in all disciplines—many of whom, however, may not be prepared for this new responsibility.

I grew up in a working-class family in a decidedly uneducated, working-class community. At school, I was lucky enough to have my eyes opened to the value of education. As a result, I have been able to lead a life that, at least in my eyes, is far richer and more fulfilling than that of most of my childhood contemporaries. I would like to see others have the same opportunities I had.

To my mind, people who make restrictive choices will inevitably lead poorer lives. Quantitative literacy and a basic understanding of science enfranchise people just as much as basic literacy. I think we help people lead richer lives if we develop their appreciation of films, plays, novels, poetry, and music. The same is true if we develop their ability to appreciate and understand quantitative data and the patterns of mathematics. The individual who can appreciate the beauty and power of literature *and* of mathematics is better off than a person who can only appreciate one, whichever one it is. We don't have to make a choice, one or the other. We can have both.

—*Keith Devlin*

As is evident from the diverse views of quantitative literacy in this volume, people perceive and use mathematics in a variety of ways. In much the same way that English is used differently in different situations—when we write, when we speak with colleagues, when we talk with friends and family—people use mathematics in different ways. "Back-of-the-envelope" calculations or approximations may be fine when computing the tip and dividing a restaurant bill among friends, but we calculate more carefully when preparing income taxes; an approximation of how much paint is needed to cover the walls of a room is fine, but measurements for new vertical blinds must be done with more precision.

Among the many people with a stake in quantitative literacy are mathematicians and mathematics educators. To gain some insight into what these experts think, nine mathematicians and mathematics educators (see sidebar) were asked via e-mail about some of the issues raised by contributors to this volume. These mathematicians represent a broad range of institutions and each brings to the discussion a personal perspective and expertise in a particular area of mathematics or mathematics education. Because of their depth of training and high expectations, mathematicians might be expected to define quantitative literacy differently from people in other professions. It turns out, however, that quantitative literacy is not a subject to which mathematicians have given much careful thought.

The interviewees are probably typical of most mathematicians in that they find it easier to discuss topics more closely related to mathematics itself and harder to formulate a concrete definition of quantitative literacy. Not surprisingly, many of their comments focus on school mathematics, and they frequently retreat to the familiar territory of naming topics that should be included in the curriculum. At the same time, they make a conscious distinction between quantitative literacy and the more comfortable area of mathematical literacy. Generally, these mathematicians argue that while all citizens need to be quantitatively literate, only some need to be mathematically literate. They believe all teachers should share the responsibility for helping students become quantitatively literate, but recognize that mathematics teachers must play a special role in that endeavor.

Mathematicians Talk About Quantitative Literacy

Mathematicians who teach at the high school, college, or university level face a dilemma. For the most part, their training was in "pure mathematics," mathematics for the sake of learning more mathematics. Applications to "real-world" problems were not at issue because that was not the purpose of their study. For them, the use of mathematics outside the classroom—to interpret data, handle personal finances, or undertake household construction projects—is effortless, something they do without much thought or angst. They acquired their quantitative literacy as part of their training in mathematics but, according to their comments, it is something they have not had much cause to think deeply about or try to define in specific terms.

When asked what constitutes quantitative literacy, the interviewees tend to think in academic terms—what mathematics should be part of the school

The Interviewees

Gail Burrill, a former high school mathematics teacher, is currently president of the National Council of Teachers of Mathematics and an associate at the National Center for Research in Mathematical Sciences Education at the University of Wisconsin.

Keith Devlin is dean of science at Saint Mary's College of California and editor of *Focus,* the newsletter of the Mathematical Association of America.

Carole Lacampagne is director of the National Institute on Postsecondary Education, Libraries, and Lifelong Learning of the U.S. Department of Education's Office of Educational Research and Improvement. Previously she was a professor of mathematics at Northern Illinois University.

William James (Jim) Lewis is chair of the mathematics department at the University of Nebraska at Lincoln and chair of the Science Policy Committee of the American Mathematical Society.

Carolyn Mahoney is Founding Faculty Professor of Mathematics at California State University San Marcos. She has also served as a program director in the Office of Systemic Reform at the National Science Foundation.

Jack Price is co-director of the Center for Mathematics and Science Education at California State Polytechnic Institute at Pomona. A past president of the National Council of Teachers of Mathematics, Price previously was superintendent of schools in Palos Verdes, California.

Sharon Ross is professor of mathematics at DeKalb College in Atlanta. She is a past vice president of the Mathematical Association of America and former editor of *AMATYC News,* the newsletter of the American Mathematical Association of Two Year Colleges.

Paul Trafton is professor of mathematics education at the University of Northern Iowa and a member of the Board of Directors of the National Council of Teachers of Mathematics.

Zalman Usiskin is professor of mathematics education at the University of Chicago and director of the University of Chicago School Mathematics Project.

curriculum? At the same time, they realize that they often lapse into a definition of *mathematical* rather than *quantitative* literacy. The general consensus is that quantitative literacy is a subset of mathematical literacy, the latter being what is required for advanced study or for mathematics- or science-related careers. "More than anything else, I think that to be quantita-

How's Your Quantitative Literacy?

What would an essay about mathematicians be without a mathematics problem? Here are some tests of quantitative literacy suggested by our interviewees:

- Some years ago, a U.S. president boasted that the rate of increase in inflation had fallen. Was this something to boast about?
- What besides grades influences a student's grade-point average? What besides prices influences the Consumer Price Index? How reliable are indices like these that involve weighted averages?
- How can you make a fair price comparison between a 19-inch, a 25-inch, and a 27-inch television set?
- If you charge $100 per month for two years to a credit card that charges 1.35 percent interest per month, but make only the minimum monthly payments of 2 percent of the outstanding balance, at the end of the two years how much will you have paid and how much will you still owe?

tively literate, one must have 'number sense' and the ability to reason with numbers," says Jim Lewis, chair of the mathematics department at the University of Nebraska at Lincoln. "I see quantitative literacy as the first level of being a mathematically literate citizen—having a big-picture view of how to work with numbers, relationships, or patterns. To me, mathematical literacy represents a higher level of mathematical competence, perhaps involving the ability to use mathematics in one's work."

Carole Lacampagne, a senior official at the U.S. Department of Education, extends Lewis's point by suggesting that there may be levels of quantitative literacy. "I see quantitative literacy as being at the level of reading, interpreting, and making simple applications—not as deep as mathematical literacy. Applying the Pythagorean Theorem should be part of a low-level quantitative literacy, while applying the Binomial Theorem should be part of a higher-level quantitative literacy. It's not a simple matter of being quantitatively literate or not. There are levels of such literacy."

"I think quantitative literacy and mathematical literacy are completely separate," counters Keith Devlin, dean of science at Saint Mary's College of California. "In addition to basic quantitative literacy, I think we prepare our children to lead fuller lives if we develop their awareness of some of the major ideas of science and mathematics—calculus, physics, biology, chemistry. But just as there's no reason to expect everyone to know how to play a musical instrument or write a screenplay, we would not expect all people to develop a deep understanding of complex mathematics. Moreover, I think mathematical literacy mostly involves qualitative issues, not quantitative ones."

The notion of separating qualitative and quantitative thinking is reinforced by Jack Price, professor of mathematics at California State Polytechnic Institute and past president of the National Council of Teachers of Mathematics (NCTM). "To me, quantitative literacy implies numbers and a basic understanding of operations on rational numbers, essentially an eighth-grade mathematics education. A qualitative mathematical literacy would include both basic skills and higher-order mathematical thinking, i.e., all of the other topics in the NCTM core curriculum, which should be our goal for all students."

In recent assessments of literacy, both the National Assessment of Educational Progress (NAEP)[1] and the Organization for Economic Cooperation and Development (OECD)[2] introduced a type of quantitative literacy they call document literacy, which involves reading and understanding texts containing charts, graphs, and formulas. A similar notion is evident in the interviewees' frequent references to the critical importance of the ability to read and acquire information from an increasing number of sources including books, newspapers, radio, television, and electronic media. Gail Burrill, the current president of NCTM, offers a statistician's perspective: "There is a growing need to be able to read newspaper articles and evaluate the assumptions that underlie conclusions drawn from the data presented. Data about schools, communities, and crime often influence where people choose to live, decisions they make about their jobs and their families, and how they vote. Skills that contribute to focusing a critical eye on conclusions—understanding what sampling is, being able to interpret margin of error, and learning to trust the numbers rather than making subjective judgments—are all necessary for effective citizenship."

What Should Students Study to Become Quantitatively Literate?

Children begin to develop their quantitative literacy skills at an early age—as soon as they begin to count objects and understand the relative size of portions they receive at the dinner table. (Watch what happens when a youngster is asked to share a candy bar with a sibling.) Their study of numbers is formalized when they start school, and there is an implicit expectation that students will become quantitatively literate through their study of mathematics, science, and other subjects that offer opportunities to develop and refine problem-solving skills.

Teachers, textbooks, and the curriculum help determine how well a person will be able to apply quantitative thinking and problem-solving strategies. Most of the mathematicians interviewed feel that the core curriculum advocated by the NCTM[3] provides students with the skills to become mathematically literate, but they differ as to what parts of the school curriculum constitute quantitative literacy.

Paul Trafton, professor of mathematics education at the University of Northern Iowa, suggests that elementary schools have a critical role to play in helping children develop their quantitative thinking. "Individuals' reasoning patterns reveal their understanding of quantitative ideas and their interrelationships. We know that young children frequently demonstrate remarkable mathematical intuition and inventive ways of solving problems. We also know that a curriculum that focuses on quantitative thinking in the early grades produces mature, capable thinkers who can use mathematics in a variety of ways."

A primary purpose of the mathematics program through the eighth grade is to help students develop their arithmetic skills with whole numbers, fractions, and percents and learn to use those skills along with quantitative thinking to solve problems. Beyond middle or junior high school, however, the debate is open. Are students continuing to build their quantitative literacy as they study algebra, geometry, and (sometimes) calculus? Or are they then moving beyond quantitative literacy to the more sophisticated and abstract mathematical literacy? Are there other topics that should be included in the curriculum to help future workers and voters become better qualified to fulfill their obligations?

One major change in the school curriculum in recent years is that many states now require all students to study algebra. This change is partly a result of the NCTM recommendation of a core curriculum for all students and partly because the nature of the work place is changing so dramatically. Zalman Usiskin, professor of mathematics education at the University of Chicago, is deeply committed to the goal that every student should study algebra. In an article "Why Is Algebra Important to Learn?"[4] he offers four reasons:

- Algebra is the language of generalization.
- Algebra enables a person to answer all the questions of a particular type at one time.
- Algebra is the language of relationships between quantities.
- Algebra is a language for solving certain kinds of numerical problems.

"Students should learn about formulas, exponential growth, rates of change, and algebraic descriptions of number properties," urges Usiskin. "Moreover, they should be able to work with spreadsheets that have their own algebra."

However, what appears to be a simple recommendation to implement—that all students need to learn algebra—has become complicated and controversial. Mathematicians, it seems, cannot agree on the content and pedagogy of high school algebra. Most agree with Usiskin that there are important ideas that can best be communicated by using the symbols of algebra, but they disagree about the usefulness of some topics in the traditional curriculum. The Education Department's Lacampagne, who in 1993 organized a conference on the role of algebra in education, suggests that quantitative literacy may require a different view of the traditional curriculum. "Students certainly should be required to learn some algebra. But I think I'd limit it to being able to read, interpret, and apply common formulae and to being able to translate words into equations to create simple mathematical models." These are the kinds of algebraic skills students will need in their careers and in their lives.

Opinions about the role of geometry are equally varied. For some, geometry is part of quantitative literacy; for others, it is part of mathematical literacy but beyond the realm of quantitative literacy. Once again Usiskin offers some interesting insights:

"One must consider two directions when examining the role of geometry in quantitative literacy. First, there are situations that are intrinsically numerical where we find spatial representation to be helpful. This covers situations such as when we represent the results of a poll in a pie chart or bar graph, or represent time-series data with a coordinate graph. There are some who worry about these representations—who think we are doing too much with the graphs of functions at the expense of dealing with their algebraic formulas. But I think most would view this geometric representation of arithmetic or algebraic ideas as one avenue to understanding those ideas—as a means to an end.

"The second direction begins with geometry, that is, with spatial questions such as the optimal shape for soap bubbles, or which figures will tessellate, or the design of a curve of constant width. To these, the phrase 'quantitative literacy' does not seem to apply even in a broad sense. We could also argue that length and area do not begin as arithmetic; they are physical geometric ideas. Being literate about them is as much geometric as it is quantitative."

For decades the two-column proofs taught in Euclidean geometry were touted as a way to help students learn to make logical arguments. However, experience has shown that this strategy generally does not work. Another option for teaching students how to reason effectively is to offer courses in logic. Once again, the result has often not been what educators expected. Thus we are still faced with the problem of how to help students learn to think logically.

While formal logic is often thought of as part of philosophy, for those such as logician Devlin, thinking logically is part of quantitative literacy. "I think the ability to reason is something that should be developed in all our classes, whether it's solving a problem in mathematics, analyzing a novel in English, making deductions from experimental results in science, or arguing a position in history or social studies. Knowing how to construct and recognize a sound argument is surely an important skill that can be acquired with training and practice in any number of domains."

NCTM president Burrill emphasizes the importance of statistics in understanding logical reasoning. "To be quantitatively literate, people don't need to learn the laws of logic in a formal sense, but they should understand what a conditional is and the difference between 'and' and 'or.' They also need a basic understanding of statistics, data analysis, and probability. The ability to draw conclusions based on statistical reasoning is critical. To do this, people need a well-developed sense of how numbers (e.g., percents, rates) are derived from real situations such as driving records, accidents, and population density. The ability to read data and draw conclusions from two-way tables is another important skill.

"A careful study of variability and how to quantify it," continues Burrill, "can lead to detailed use of quality control charts as well as provide researchers with the tools they need when deciding whether a drug is safe to market or not. A sense of order in random behavior is necessary for people in any profession if they are to organize and process information. The use of graphic displays to sort information, as well as to search for patterns and relationships, is critical even to those who write newspaper articles. These skills will be needed for any job from psychologist to factory worker to clothing designer."

Other interviewees point out ways in which more advanced topics in mathematics are important for full understanding of ideas in other disciplines. Sharon Ross, professor of mathematics at DeKalb College, suggests including in mathematics education areas often unfamiliar to nonmathematicians. "I'd like to see more done with voting, apportionment, fair division, and game theory to show students the connections of mathematics with the social sciences." And Nebraska's Lewis seems to be thinking about business and technology when he says, "Some elementary numerical analysis is important (significant digits, round-off error) to provide a solid ability to work with ratios, percentages, and order-of-magnitude estimations (e.g., how big the deficit is compared to the GNP)."

How Will Students Become Quantitatively Literate?

Helping students become quantitatively literate, the interviewees agreed, should be a shared academic responsibility. Mathematics departments cannot do it alone; they will have to work with teachers in other disciplines. "Many teachers can make very significant contributions. Most can emphasize quantitative ideas more than they do now. All can advance the idea that some problems are solved collectively with everyone bringing a different strength to the problem," says Lewis. "In my classes, I sometimes venture into science issues I don't fully understand. My job is not to be the science teacher but to help add some mathematics to a situation, allowing someone else to add depth of understanding of the science issues. All teachers should do the same with quantitative ideas."

Yet enthusiasm for this idea of shared responsibility is dampened by concern about the difficulties of having people not trained in mathematics integrate mathematics into their classes and present mathematical concepts to students. All teachers are expected to contribute to the development of students' verbal literacy skills. Not surprisingly, however, mathematicians express serious doubts as to how well-equipped elementary school teachers or teachers of other secondary school subjects are to meet the similar challenge of helping students become quantitatively literate. "We cannot assume that all elementary school teachers are quantitatively literate," says Usiskin. "Many of them purposely do not teach beyond a certain grade because they are afraid of the mathematics or the science they would need to discuss at the higher grade."

But teachers should not avoid their responsibility to help students develop quantitative skills. "Quantitative literacy will soon be like reading—a tool necessary to function in any field," asserts Burrill. "All teachers should be able to apply quantitative techniques to the vast amount of information they encounter in their own field." Suggestions for building teachers' mathematical abilities and confidence focus on initiating collaborative efforts between mathematics teachers and others. "Currently, the responsibility seems to rest on the mathematics and science teachers," Burrill continues. "It is important that we bring the entire education system into the fold so that each discipline reinforces the ability to organize and use information properly."

Carolyn Mahoney, a former program officer at the National Science Foundation, examines ways in which collaborative efforts can be initiated at the local level. "The mathematical community must spearhead the effort to help other teachers incorporate quantitative literacy skills into their classes. College and university mathematics faculty can work with their colleagues in teacher education to improve the mathematical preparedness of students intending to teach. Regional and local mathematics organizations have a responsibility to work with superintendents, principals, and teachers to create resources and infrastructures that support improved knowledge acquisition, school restructuring, and other needed changes. Classroom teachers must be provided time and other support to create and implement curricular and structural changes. And mathematics teachers must help other teachers better understand the mathematical content relevant to other subjects, as indeed other teachers must help mathematics teachers with unfamiliar disciplinary content. Some recent efforts at incorporating 'Mathematics Across the Curriculum' are exemplary, although most are occurring at the college level. In such programs, individual responsibilities are different but the overall responsibility is shared."

Setting Mathematics in Context

If the goal of quantitative literacy is to enable people to function in a society in which quantitative and analytic skills are growing in importance, then those skills will be better learned in contexts that are meaningful to students. In high school, that means introducing quantitative thinking in science class, in social studies class, in economics class, and in history class. As DeKalb's Ross aptly puts it, "Students need experiences where they develop and apply mathematics so they gain confidence in their skills. We have to put an end to the isolation of the mathematics classroom from the rest of the school and from the rest of the world. I'm not saying assign only work place problems or applications, but whatever students are interested in should be studied from a mathematical point of view."

"The fact is, surely, we forget anything we do not use on a regular basis," Devlin adds. "In contrast, we remember things we make regular use of. So I think it's not so much a question of what students study, as how we equip them to lead a life that involves regular use of certain skills. A person who is able to make sense of a newspaper article that involves quantitative information, graphs, and charts and who sees the value of being able to understand such an article will surely continue to use those basic quantitative literacy skills. In contrast, people who do not see any value in reading such articles will simply never bother to read them—they will 'turn off' at the first whiff of anything vaguely quantitative. And no amount of drill and practice in the school classroom will have a lasting effect; what is learned will simply be forgotten."

The value of providing opportunities for students to use mathematics in a variety of subjects is not simply to enable them to learn specific concepts but to illustrate that mathematics is embedded in many contexts. It takes special skills to uncover how quantitative thinking can be used in a "real-life" situation in which there is a problem to be solved. Again, Mahoney points out the special role mathematics teachers play both in teaching mathematics in context to their own students and in helping other teachers use quantitative skills and reasoning to enrich their courses. "Mathematics teachers have a responsibility to incorporate contexts and applications in the mathematics classroom. I believe they have an added responsibility to provide leadership so all teachers know where and how to incorporate applications of mathematics into their courses. The ability to recognize what mathematics to use and the ability to use it correctly is critical. In my experience, students become engaged in learning through connections to their daily lives. They remember what they learn longer when actively engaged in the learning process and they stay engaged when teacher-guided opportunities of sufficient conceptual depth encourage reflection, abstraction, and pattern recognition."

Price, himself a former vocational education teacher, emphasizes the fact that learning mathematics in context also helps students acquire valuable tools they need for lifelong learning. "I am not certain that any specific mathematics we teach in high school will enable students to remember enough to be quantitatively literate. Even when we taught 'consumer math' it was so far removed from the time of actual application by the students that it was nearly worthless. We need continually to bring in real-life problems so students learn to solve them and develop the tools they will need when faced with other problems later in their lives. It is the process they need to learn rather than specific content."

> As someone who came to the United States from Europe as an adult, I have often observed that here it is socially acceptable (and sometimes appears to be an absurd matter of pride) to be quantitatively illiterate. Americans who cannot read are generally ashamed of that fact; yet countless Americans will cheerfully declare that they cannot add up or make sense of quantitative data. I don't see how the education system can rectify the problem of quantitative illiteracy without there first being a change in social attitudes. It's surely one of the most striking social paradoxes that the nation that leads the world in science and technology, and in which most of its citizens benefit from the fruits of its scientific expertise, is the one that places the least value on such skills.
>
> So where should we be putting our greatest efforts? Into changing the public perceptions of quantitative literacy and of mathematics. Until we make progress in that area, we are unlikely to succeed in anything else in terms of quantitative literacy and mathematics education for the population at large.
>
> —*Keith Devlin*

Endnotes

1. I. S. Kirsch, A. Jungeblut, L. Jenkins, and A. Kolstad, *Adult Literacy in America* (Washington, D. C.: National Center for Education Statistics, 1993).

2. Organization for Economic Cooperation and Development, *Literacy, Economy, and Society* (Ottawa, Canada: Statistics Canada 1995).
3. National Council of Teachers of Mathematics, *Curriculum and Evaluation Standards for School Mathematics* (Reston, Va.: National Council of Teachers of Mathematics, 1989).
4. Zalman Usiskin, "Why Is Algebra Important to Learn?" *American Educator* (Spring 1995) 30–37.

Appendix: Defining and Measuring Quantitative Literacy

JOHN A. DOSSEY
Illinois State University

As the essays in this volume were being written, the College Board's Mathematical Sciences Advisory Committee undertook a separate study of how quantitative literacy is reflected in a variety of common national examinations. Although this study is still a work in progress, many readers of this volume may be interested in a preliminary report. With this in mind, we include the following summary prepared by committee chair John Dossey.

Creating a working definition of quantitative literacy is an important task, given the significant position occupied by the mathematical sciences in today's world. However, there is no widely accepted definition of this term. The purpose of this study is to establish such a definition and to suggest how we might assess students' acquisition of aspects of quantitative literacy in various curricular domains.

To define and measure quantitative literacy, we need a model that spans a wide variety of contexts. Historically, definitions have been expressed in terms of bodies of knowledge such as arithmetic, algebra, geometry, trigonometry, analytical geometry, and calculus. This approach no longer suffices, even though the study of these branches of mathematics is necessary to understand our world. To understand the meaning of quantitative literacy, a better model is one based on a categorization of mathematical behaviors into six major aspects:

- Data representation and interpretation
- Number and operation sense
- Measurement
- Variables and relations
- Geometric shapes and spatial visualization

- Chance

These aspects provide a broad basis for examining the ability to interpret and act in a wide variety of mathematics-related settings.

Quantitative literacy may be defined as the ability to interpret and apply these aspects of mathematics to fruitfully understand, predict, and control relevant factors in a variety of contexts. As here defined, quantitative literacy constitutes a significant part of the gateway to advanced mathematics. It is grounded in a strong command of school mathematics through the equivalent of two years of algebra and one year of geometry, with broad coverage of other core areas as outlined in the National Council of Teachers of Mathematics' *Curriculum and Evaluation Standards for School Mathematics.*[1]

In this definition of quantitative literacy, it is important to note the role of the "taught curriculum" in mathematics that quantitatively literate people might be expected to have completed as part of their formal schooling. The expectation of the equivalent of two years of algebra and one of geometry would equate to an education that included the formation and solving of linear, quadratic, and systems of equations; understanding exponential growth; the ability to create and interpret graphs; knowing the basic concepts and relationships of elementary geometry, including the Pythagorean Theorem and some trigonometry of the right triangle; and an introduction to concepts and skills related to probability and statistics.

This model for quantitative literacy is probably best illustrated through a set of exemplars depicting the types of tasks a quantitatively literate person could be expected to successfully perform. The cognitive demands such tasks place on people are determined by the interaction of a multitude of factors. Foremost among these is the degree of understanding of concepts and the degree of ability to apply procedures. Both conceptual and procedural knowledge are needed in problem solving; the former as a base for analyzing and synthesizing information, the latter for knowing how and when to apply algorithmic or heuristic methods to derive solutions.

One approach to these levels of understanding is outlined in Table 1. Despite appearances, these levels are not discrete categories in which people or tasks can be placed, but rather guides to help our thinking about degrees of quantitative literacy. They are but propadeutics for understanding the contextual complexity of settings calling for the application of quantitative knowledge and skills. The levels are illustrated below by specific tasks, each of which is assigned a level of cognitive demand. These assignments are intended primarily to aid in understanding the levels, not to suggest that such distinctions can be made on a task-by-task basis.

The relation of these levels to the complexity of tasks can be seen in the following diagram, where the rows identify the difficulty of the underlying context, the columns identify the degree of novelty, and the numerals refer to levels of understanding.

	Routine	*Novel*
Basic	1	3
Advanced	2	4

Although this characterization is really continuous, as are the levels themselves, this discrete model helps reveal the relation of levels to tasks. It offers a simple way of thinking about the criteria by which we might describe the nature of quantitative literacy.

Table 1.
Levels of Conceptual and Procedural Understanding.

Conceptual Understanding	*Procedural Understanding*
Level 1: Recognize and apply basic concepts in routine settings; use representations (verbal, graphic, tabular, or symbolic) to illustrate simple concepts and relationships.	*Level 1:* Perform routine calculations; apply common formulas; solve simple equations.
Level 2: Recognize and apply more advanced concepts related to a given aspect; determine equivalences among numerical quantities; write equations; choose and use appropriate representations for exemplars of concepts and relationships; reason inductively; determine basic spatial properties and relationships; understand when data are sufficient for decision making.	*Level 2:* Select algorithms appropriate to familiar problems and apply them accurately; round and order numbers as appropriate to routine problems; solve more complex equations; describe the nature of the answer needed for a given problem (units, size).
Level 3: Develop simple models for contexts involving rates and proportion; apply models in simple or familiar contexts; translate among representations; reason deductively; justify conclusions informally.	*Level 3:* Extend and modify algorithms and procedures to fit specific contextual needs; justify the appropriateness of a given procedural approach; deal effectively with issues of dimensional analysis in more advanced settings.
Level 4: Develop and apply complex models in novel contexts; extend or modify representations to fit advanced and novel settings; engage in more complex chains of deductive reasoning; write a paragraph describing a nonroutine problem and detailing the solution.	*Level 4:* Develop new algorithms or substantially modify known ones; describe where algorithms apply and where they do not.

Data Representation and Interpretation

Data representation and interpretation are perhaps the most basic aspects of quantitative literacy because they are the aspects through which people perceive data, collect information, and construct models for decision making in quantitative settings. That the skills of collecting, organizing, reading, representing, and interpreting data are important is reflected in the ubiquitous nature of such skills in a person's daily life. These understandings and skills run the gamut from the ability to find a desired piece of information in a chart to the ability to describe the role variability plays in analyzing a set of data. For example, here is a level 1 task:[2]

> You wish to use the automatic teller machine at your bank to make a deposit (see Figure 1). Figure the total amount of the two checks being deposited and enter the amount on the form in the space next to "TOTAL."

People who are quantitatively literate can, given a task with associated data, construct simple frequency tables, pie charts, or bar graphs; compute the mean, median, mode, and range; and sort out data that are relevant to the task from those that are not. They can also construct and read scatter-plots

Availability of Deposits

Funds from deposits may not be available for immediate withdrawal. Please refer to your institution's rules governing funds availability for details.

Crediting of deposits and payments is subject to verification and collection of actual amounts deposited or paid in accordance with the rules and regulations of your financial institution.

PLEASE PRINT

YOUR MAC CARD NUMBER (No PINs PLEASE)
111 222 333 4

YOUR FINANCIAL INSTITUTION
Union Bank

YOUR ACCOUNT NUMBER
987 555 674

YOUR NAME
Chris Jones

CHECK ONE ☐ DEPOSIT or ☐ PAYMENT

CASH	$	00
LIST CHECKS BY BANK NO.	ENDORSE WITH NAME & ACCOUNT NUMBER	
	557	19
	75	00
TOTAL		

DO NOT DETACH TICKET

DO NOT FOLD NO COINS OR PAPER CLIPS PLEASE

Figure 1.

and box-plots and draw simple inferences (e.g., comparisons) from such representations. At the same time, they can follow directions requiring branching (e.g., flowcharts) or read information from complex tables giving schedules or describing alternative choices. The task below illustrates a level 3 understanding of data interpretation:

> The graph in Figure 2 gives the rewind time in seconds for cassette tapes versus the tape length in minutes. Which of the following best describes the relationship shown?
> (a) As the tape length increases, the rewind time decreases.
> (b) As the tape length increases by 5 minutes, the rewind time decreases by 10 seconds.
> (c) As the tape length increases by 5 minutes, the rewind time increases by 10 seconds.
> (d) As the tape length decreases by 5 minutes, the rewind time increases by 10 seconds.
> (e) As the tape length increases by 10 minutes, the rewind time increases by 10 seconds.

At the upper levels of performance in data representation and interpretation, people can succeed at tasks that require higher-level inferences and responses to various data sources. They can use measures of central tendency and variability to compare and contrast different groups of data; they can construct an appropriate representation for a set of data, conduct rel-

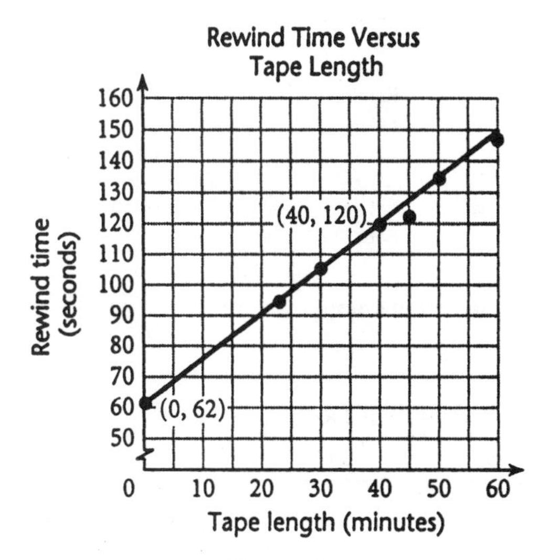

Figure 2.

evant analyses (e.g., trend and strength of relationship), and communicate cogent arguments based on their representations and analyses. They can also effectively evaluate graphical information for bias and subtle misrepresentation, as well as view the role variability plays in such situations. This level of understanding is illustrated by the following level 4 task:[3]

> You work for a business that has been using two taxicab companies, Company A and Company B. Your boss gives you a list (see Figure 3) of (early and late) "arrival times" for taxicabs from both companies over the past month. Your job is to analyze these data using charts, diagrams, graphs, or whatever seems best. You are to (i) make the best argument that you can in favor of Company A; (ii) make the best argument that you can in favor of Company B; and (iii) write a memorandum to your boss that makes a reasoned case for choosing one company or the other.

Company A		*Company B*	
3 mins 30 secs	Early	3 mins 45 secs	Late
45 secs	Late	4 mins 30 secs	Late
1 min 30 secs	Late	3 mins	Late
4 mins 30 secs	Late	5 mins	Late
45 secs	Early	2 mins 15 secs	Late
2 mins 30 secs	Early	2 mins 30 secs	Late
4 mins 45 secs	Late	1 min 15 secs	Late
2 mins 45 secs	Late	45 secs	Late
30 secs	Late	3 mins	Late
1 min 30 secs	Early	30 secs	Early
2 mins 15 secs	Late	1 min 30 secs	Late
9 mins 15 secs	Late	3 mins 30 secs	Late
3 mins 30 secs	Late	6 mins	Late
1 min 15 secs	Late	4 mins 30 secs	Late
30 secs	Early	5 mins 30 secs	Late
2 mins 30 secs	Late	2 mins 30 secs	Late
30 secs	Late	4 mins 15 secs	Late
7 mins 15 secs	Late	2 mins 45 secs	Late
5 mins 30 secs	Late	3 mins 45 secs	Late
3 mins	Late	4 mins 45 secs	Late

Figure 3.
Arrival Times for Company A and Company B Taxicabs.

Number and Operation Sense

Number and operation sense encompasses basic number systems, the representations of numbers within these systems, and the operations characterizing the systems of whole numbers, integers, rational numbers, and real numbers. Performance in this aspect of mathematical behavior includes the ability to carry out one- and two-step operations involving comparisons, ratios, and percentages; to address questions of relative size and equivalent forms of numbers; and to use numbers to describe attributes of objects from the external world.

In practice, quantitatively literate people understand when a given operation is appropriate and are able to complete the necessary additions, subtractions, multiplications, and divisions. Such understanding is illustrated by the following level 2 task:[4]

> Your neighbor hired you and two of your friends to rake leaves. The house has a back yard and a front yard that are about the same size. The neighbor agreed to pay the three of you $60 for the entire job. On the day of the job, you and one friend arrived to start the job at 9 a.m. By the time your second friend came, the front yard was finished. All three of you then finished the back yard together. How should the money be split among you?

Beyond this, quantitatively literate people are capable of applying the principles of basic number theory to describe relationships between numbers and to compute both greatest common factors and least common multiples. Such people can complete basic computations and comparisons both mentally and through traditional means involving both paper-and-pencil and concrete model representations. More complex computations should be done using technology such as calculators and spreadsheets. In each case, people should be able to provide estimates of the relative magnitude of the expected answer using rounding, scientific notation, and comparisons as appropriate.

At the upper end of performance in number and operation sense, people are able to employ basic operations and number comparisons in situations where they must assemble data from varied sources and analyze those data using inferred operations and models. They can solve problems requiring direct, inverse, and joint proportional reasoning; estimate rates of change; provide a rationale for the selection of data and level of precision required by the operations and models they use; examine alternative algorithms, showing why they work or in which cases they fail; develop models for problems involving real-world data; communicate the basis for their models; and describe both the validity and reliability that might be associated with "solutions" drawn from their models. This level of understanding is illustrated by the following level 4 comparison item:[5]

> The Peterson family rents 30 videotapes yearly, of which 23 are rented for one night only and 7 are rented over a period of two nights.

Video Store A: \$2.65 per tape for one night; \$1.50 charge for each additional night; every tenth tape free for one night
Video Store B: \$3.00 per tape for two nights; one credit if tape returned after one night; every 10 credits = one free rental

What would the total yearly cost be for the Petersons at each store?

Measurement

Measurement plays a very important role in the everyday quantitative actions of citizens. It is perhaps the most visible, but least considered, aspect of quantitative literacy, covering the range from common linear measurement to the application of mathematics to derived quantities describing economic indices (GNP, person-hours) or mechanical units (ergs, joules).

At the basic levels, measurement involves recognition, selection, and application of appropriate units of measure; selection and use of basic measurement tools (e.g., ruler and protractor); estimation of linear measurements, and conversion of measures within the traditional system. People performing at these basic levels can use perimeter, area, and volume formulas to find appropriate measures of basic figures (square, rectangle, cube, parallelepiped, circle, cylinder, and triangle). For example, here is a level 2 item:[6]

> A circle with center O and radius of length 3 units is inscribed in a square (see Figure 4). What is the area, in square units, of the shaded region?

As people progress in their ability to deal effectively with measurement situations, they can develop derived measures using simple measures, as well as interpret common complex units derived from simple units of measurement (miles/hour, person-hours, blood alcohol levels, utility-rate-use units). They are able to find the areas of irregular figures when given the relevant information.

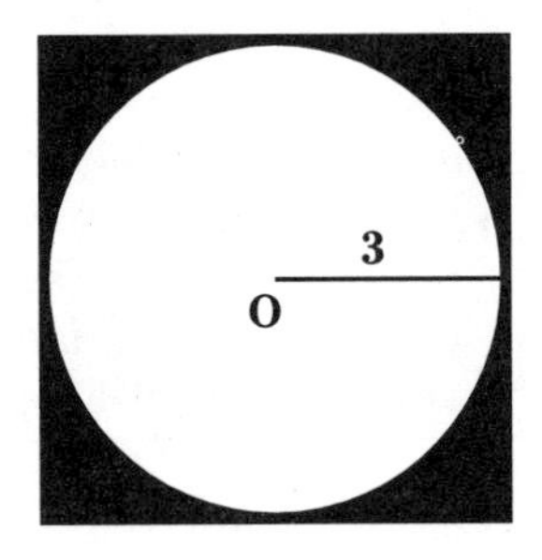

Figure 4.

At the upper levels of performance in measurement situations, people are able to recognize when perimeter, area, or volume is invariant under specific transformations of a region or solid. They understand the importance of the precision of a measure and interpret it in terms of significant digits and tolerances. When faced with a novel problem, they can decide upon a system of measurement, select appropriate units (e.g., develop a scale for substance hardness), carry out the required measurements, connect such units to the original applied situation, and communicate the meaning of their measurements. The following level 4 task presents a problem that draws on both geometric and measurement knowledge in product design:[7]

> Explain how a straight straw that is attached to the side of a rectangular juice box can still be long enough that it will not fall into the box when it is inserted into the predesigned hole in the top of the box. (See Figure 5.) Where should the predesigned hole be placed to assure that as much as possible of the straw is guaranteed to stick out of the top of the box?

Variables and Relations

The concepts and skills associated with the aspect of variables and relations reflect modes of representation and procedures often associated with algebra. However, they go beyond symbolic manipulation to the ability to repre-

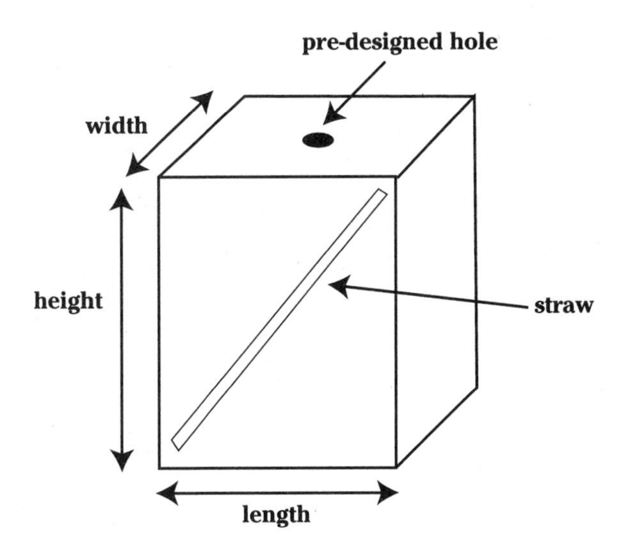

Figure 5.

sent change by means of tables, graphs, symbols, and words. This aspect also addresses a person's ability to make sense of formulas, to understand general properties of mathematical systems, and to use this information in powerful ways to represent and solve problems involving unknown quantities.

People at the lower levels of performance in variables and relations are able to identify the values that different quantities might assume. They can recognize and extend patterns involving numerical or geometric content. They can evaluate a simple formula in a context, given both the formula and the values to be substituted into the formula. A level 2 task is illustrated below:[8]

> The cost to rent a motorbike is given by the following formula:
>
> $$\text{Cost} = (\$3 \times \text{number of hours}) + \$2$$
>
> Fill in the missing values in the table below:

Time in Hours	*Cost in Dollars*
1	5
4	
	17

At the middle levels of performance, people will be able to represent and interpret situations through the use of tables, equations involving variables, or graphs of relationships. They can relate values in a situation to a linear or quadratic equation, represent such equations graphically, and solve for required values. They can use rates as quantities (miles/hour, person-days) and can reason about quantities in complex ways.

At the upper levels of performance in variables and relations, people can express relationships in functional form, manipulate that form symbolically, and transform the form to tabular or graphical representations when needed. They can solve problems involving relations and operations concerning rates and other derived variables (average speed, GNP). The problem below illustrates level 4 quantitative literacy in variables and relations:[9]

> Figure 6 shows a single shopping cart as well as 12 shopping carts that have been nested together. The drawings are 1/24 the real size. Create a rule that will tell you the length S of storage space needed for the carts when you know the number N of shopping carts to be stored.

Geometric Shapes and Spatial Visualization

The ability to relate geometric concepts to spatial settings in a person's environment is an important aspect of quantitative literacy. The relation-

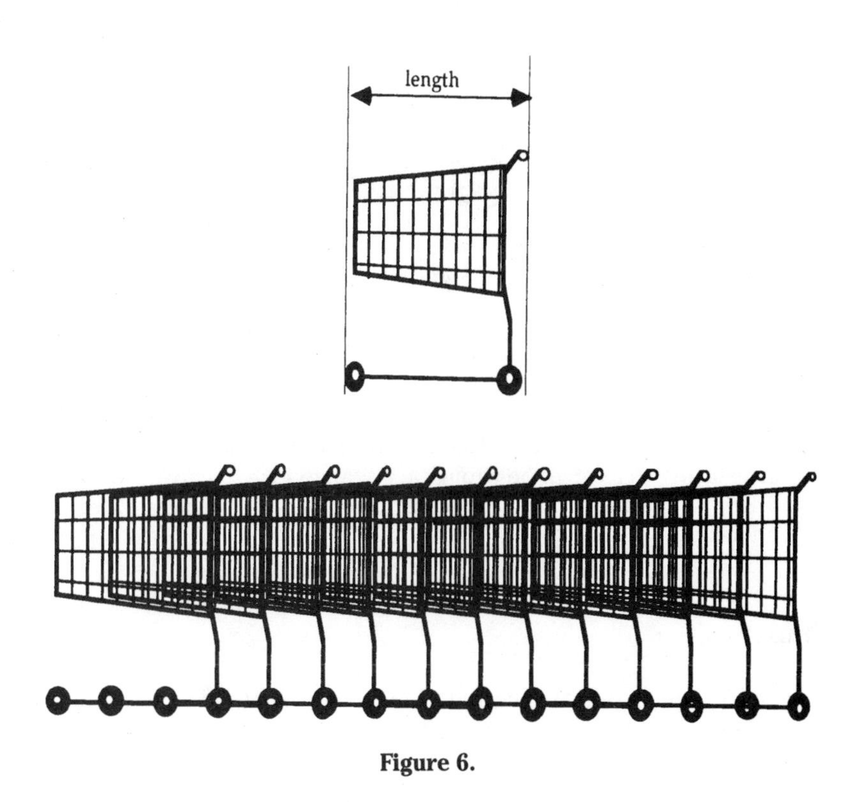

Figure 6.

ship of this aspect to problem solving extends far beyond the ability to identify, name, and compare geometric figures in two and three dimensions. At the lowest level, people can communicate spatial relationships through sketching, description, and other means. A level 2 understanding is reflected in the following task:[10]

> Which of the six figures in Figure 7 could be exactly the same shape?

Using geometric knowledge requires that people analyze shapes and identify their basic characteristics and properties (e.g., alternate interior angles for parallel lines or the Pythagorean Theorem in the form $a^2 + b^2 = c^2$). At a higher level, the quantitatively literate person can mentally transform shapes or find their projections in two or three dimensions. At the highest level of quantitative literacy, people are capable of making inferences from geometric representations and selecting appropriate representations for describing a situation or relationship. Such people can translate between synthetic representations of a geometric situation and coordinate-based represen-

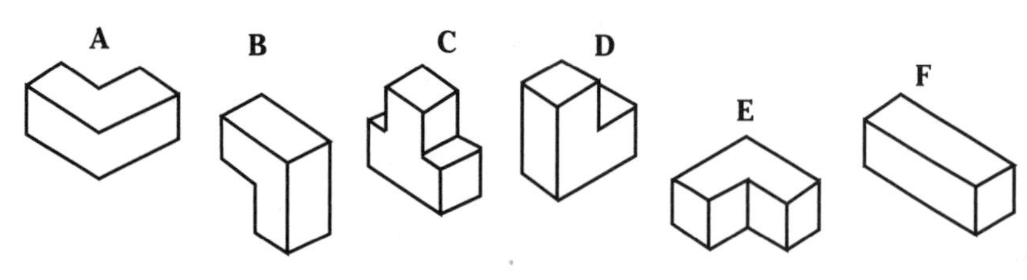

Figure 7.

tations, and use both in the solution of spatial problems. A level 3 task is illustrated below:[11]

> Examine the plane figures in Figure 8, some of which are not drawn to scale. Investigate what might be wrong (if anything) with the given information. Briefly write your findings and justify your ideas on the basis of geometric principles.

Chance

Dealing with uncertainty is common in the application of mathematical knowledge to situations of daily life. These situations range from the inter-

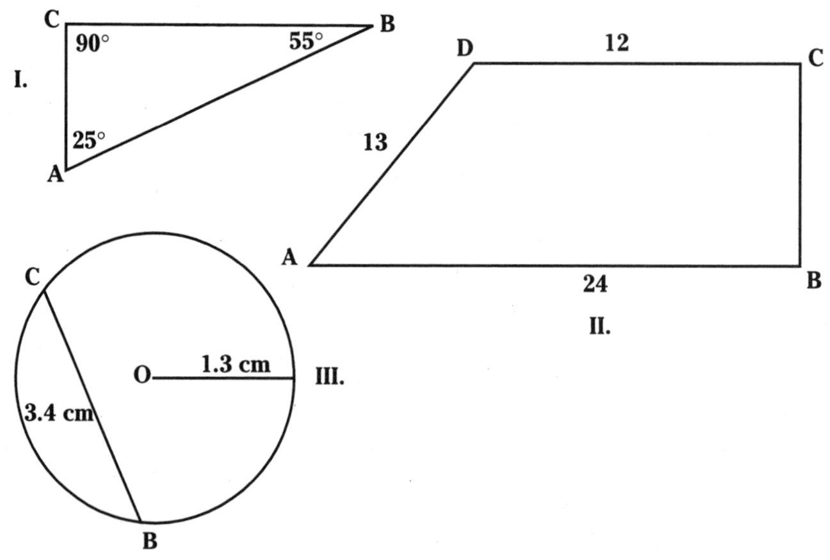

Figure 8.

pretation of weather forecasts to the consideration of odds in a lottery. Quantitative literacy begins with the application of simple probability models to discrete sample spaces. People at this low level of performance are able to identify intuitively which of two options is more likely; to apply probability concepts and relations to compound events; and to recognize dependence and independence in the context of a real-world situation. A level 2 understanding of chance is captured by the following task:[12]

> An advertising firm reports that a certain advertisement on regional radio and television is heard by 30 percent of the regional population on radio and 20 percent of the population on TV (see Figure 9). (Note that 25 percent hear the ad on radio only, and 15 percent see it on TV only.)
>
> A customer walks into the store that is sponsoring the advertisement. Find the probability that the customer has: (a) seen the ad on TV; (b) seen the ad on TV *and* heard it on radio; (c) neither seen nor heard the ad; (d) either heard the ad on radio *or* seen it on TV.

The chance aspect of quantitative literacy includes the ability to match phenomena to their expected distributions in both discrete and continuous situations. At higher levels of performance, people are able to recognize and apply common statistical distributions to straightforward applied problems. They can make inferences based on probability and detect incorrect conclusions in a simple statistical modeling situation. In particular, they can correctly use probability to produce a cogent argument based on data in context and can detect incorrect inferences in more subtle statistical arguments and models. The ability called for at level 4 is illustrated by the following task:[13]

> Joe drives a minibus in his town. The bus has 8 seats. People buy tickets in advance, but on the average, 10 percent of those who buy tickets do not show up. So Joe sells 10 tickets for each trip. Sometimes more than 8 people show up with tickets. Estimate the probability that this will happen and describe the process by which you arrived at this estimate.

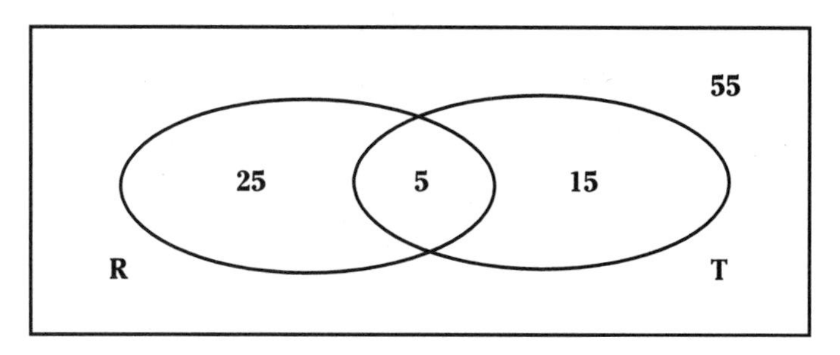

Figure 9.

Implications

People who are quantitatively literate are able to think and reason across the various aspects of mathematical behaviors, actively using concepts, principles, and skills to make sense out of situations they encounter. This requires that they integrate not only content but also the cognitive processes required to probe, interpret, conjecture, validate, and communicate what the various aspects of mathematics reveal about a given situation. The ability to use mathematics as a tool to make sense of situations in the environment requires that people model the situations (mentally or formally), bring to bear their mathematical knowledge, and work toward a solution.

Endnotes

1. National Council of Teachers of Mathematics, *Curriculum and Evaluation Standards for School Mathematics* (Reston, Va.: National Council of Teachers of Mathematics, 1989).
2. P. E. Barton and Lynn Jenkins, *Literacy and Dependency* (Princeton, N.J.: Educational Testing Service, 1995).
3. A. Schoenfeld, ed. (in process), *A Framework for Balance* (Berkeley, Calif.: Balanced Assessment Project).
4. Wisconsin Department of Public Instruction, Sample Problems for Alternative Assessment in Mathematics (unpublished manuscript, 1995).
5. J. A. Dossey, I. V. S. Mullis, and C. O. Jones, *Can Students Do Mathematical Problem Solving? Results from Constructed-Response Questions in NAEP's 1992 Mathematics Assessment* (Washington, D. C.: National Center for Education Statistics, 1993).
6. I. V. S. Mullis, J. A. Dossey, E. H. Owen, and G. W. Phillips, *NAEP 1992: Mathematics Report Card for the Nation and the States* (Washington, D. C.: National Center for Education Statistics, 1993).
7. *A Framework for Balance*
8. I. V. S. Mullis, J. A. Dossey, E. H. Owen, and G. W. Phillips, *The State of Mathematics Achievement: NAEP's 1990 Assessment of the Nation and the Trial Assessment of the States* (Washington, D. C.: National Center for Education Statistics, 1991).
9. *A Framework for Balance*
10. Quantity Project. (n.d.). Department of Mathematics Sciences, San Diego State University, San Diego, California (unpublished manuscript).
11. Ibid.
12. C. M. Newman, T. E. Obremski, and R. L. Scheaffer, *Exploring Probability* (Palo Alto, Calif.: Dale Seymour, 1987).
13. M. Gnanadesikan, R. L. Scheaffer, and J. Swift, *The Art and Techniques of Simulation* (Palo Alto, Calif.: Dale Seymour, 1987).

Recommended Reading
Mathematical and Quantitative Literacy

Geometry & Visualization

Banchoff, Thomas F. *Beyond the Third Dimension: Geometry, Computer Graphics, and Higher Dimensions.* New York, NY: Scientific American Library, 1996.

Emmer, Michele. *The Visual Mind: Art and Mathematics.* Cambridge, MA: MIT Press, 1993.

Ferguson, Claire. *Helaman Ferguson: Mathematics in Stone and Bronze.* Erie, PA: Meridian Creative Group, 1994.

Fischer, Gerd. *Mathematical Models from the Collections of Universities and Museums.* Wiesbaden, DRG: Friedrich Vieweg & Sohn, 1986.

Francis, George K. *A Topological Picturebook.* New York, NY: Springer-Verlag, 1987.

Loeb, Arthur L. *Space Structures: Their Harmony and Counterpoint.* Reading, MA: Addison-Wesley, 1976.

Loeb Arthur L. *Concepts & Images: Visual Mathematics.* Cambridge, MA: Birkhäuser Boston, 1992.

Rucker, Rudy. *The Fourth Dimension: Toward a Geometry of Higher Reality.* Boston, MA: Houghton Mifflin, 1984.

Schattschneider, Doris S. *Visions of Symmetry: Notebooks, Periodic Drawings & Related Work of M. C. Escher.* New York, NY: W. H. Freeman, 1995.

Tromba, Anthony J. *Mathematics and Optimal Form.* New York NY: Scientific American Library, 1995.

Weeks, Jeffrey R. *The Shape of Space.* New York, NY: Marcel Dekker, 1985.

Symmetry & Patterns

Grunbaum, Branko and Shephard, G.S. *Tilings and Patterns.* New York, NY: W. H. Freeman, 1987.

Huntley, H.E. *The Divine Proportion: A Study in Mathematical Beauty.* Mineola, NY: Dover, 1970.

Morrison, Philip and Morrison, Phylis. *Powers of Ten.* New York, NY: Scientific American Books, 1982.

Rosen, Joseph. *Symmetry in Science: An Introduction to the General Theory.* New York, NY: Springer-Verlag, 1995.

Stevens, Peter S. *Patterns in Nature.* Boston, MA: Little, Brown & Co., 1974.

Thompson, D'Arcy. *On Growth and Form,* (2 vols.) Cambridge, MA: Cambridge University Press, 1942; Abridged Edition, 1966.

Data Presentation & Analysis

Brook, Richard J., et al. *The Fascination of Statistics.* New York, NY: Marcel Dekker, 1986.
Burrill, Gail, et al. *Data Analysis and Statistics Across the Curriculum.* Reston, VA: National Council of Teachers of Mathematics, 1992.
Campbell, Stephen K. *Flaws and Fallacies in Statistical Thinking.* Englewood Cliffs, NJ: Prentice-Hall, 1974.
Huff, Darrell. *How to Lie with Statistics.* New York, NY: W.W. Norton & Co., 1993.
Landwehr, James and Watkins, Ann. *Exploring Data.* Palo Alto, CA: Dale Seymour Publishers, 1986.
Moore, David S. *Statistics: Concepts and Controversies,* Third Edition. New York, NY: W. H. Freeman, 1995.
Tanur, Judith M., et al. *Statistics: A Guide to the Unknown,* Third Edition. Laguna Hills, CA: Wadsworth, 1989.
Tufte, Edward R. *The Visual Display of Quantitative Information.* Cheshire, CT: Graphics Press, 1983.
Tufte, Edward R. *Envisioning Information.* Cheshire, CT: Graphics Press, 1990.
Zawojewski, Judith S., et al. *Dealing with Data and Chance.* Reston, VA: National Council of Teachers of Mathematics, 1991.

Practical Mathematics

Bolt, Brian *Mathematics Meets Technology.* Cambridge: Cambridge University Pr, 1991.
COMAP. *For All Practical Purposes: Introduction to Contemporary Mathematics,* Third Edition. New York, NY: W. H. Freeman, 1996.
Huff, Darrell. *The Complete How to Figure It.* New York, NY: W.W. Norton & Co., 1996.
Jacobs, Harold R. *Mathematics, A Human Endeavor: A Book for Those Who Think They Don't Like the Subject.* San Francisco, CA: W. H. Freeman, 1982.
Stein, Sherman K. *Strength in Numbers : Discovering the Joy and Power of Mathematics in Everyday Life.* New York, NY: John Wiley & Sons. 1996.
Tannenbaum, Peter , and Arnold, Robert. *Excursions in Modern Mathematics.* Englewood Cliffs, NJ: Prentice Hall, 1992.
Wattenberg. Frank. *Personal Mathematics and Computing: Tools for the Liberal Arts.* Cambridge, MA: MIT Press, 1990.

Numbers & Infinity

Beckman, P. *A History of Pi.* New York, NY: St. Martin's Press, 1971.
Clawson. Calvin C. *The Mathematical Traveler: Exploring the Grand History of Numbers.* Plenum Press, 1994.
Dantzig, Tobias. *Number: The Language of Science,* Fourth Edition. New York, NY: Macmillan, 1959.
Davis, Philip. *The Lore of Large Numbers.* Washington, DC: Mathematical Association of America, 1961.

Hopkins, Nigel J. *Go Figure! The Numbers You Need for Everyday Life.* Detroit, MI: Visible Ink Press, 1992.
Humez, Alexander. *Zero to Lazy Eight: The Romance of Numbers.* New York, NY: Simon & Schuster, 1993.
Ifrah, Georges. *From One to Zero: A Universal History of Numbers.* New York, NY: Viking Penguin, 1985.
Imeson, K.R. *The Magic of Number.* Leicester, England: The Mathematical Association, 1989.
Lines, Malcolm E. *Think of a Number: Ideas, Concepts, and Problems.* Philadelphia, PA: I.O.P Publishing, 1990.
Maor, Eli. *e: The Story of a Number.* Princeton, NJ: Princeton University Press, 1993.
Maor, Eli. *To Infinity and Beyond: A Cultural History of the Infinite.* Princeton, NJ: Princeton University Press, 1991.
Reid, Constance. *From Zero to Infinity: What Makes Numbers Interesting,* Fourth Edition. Washington, DC: Mathematical Association of America, 1992.
Roberts Joseph. *The Lure of the Integers.* Washington, DC: Mathematical Association of America, 1992.
Rucker, Rudy. *Infinity and the Mind: The Science and Philosophy of the Infinite.* Boston, MA: Birkhauser, 1982.
Sondheimer, Ernst and Rogerson, Alan. *Numbers and Infinity: An Historical Account of Mathematical Concepts.* New York, NY: Cambridge University Press, 1981.
Vilenkin, N. Ya. *In Search of Infinity.* Cambridge, MA: Birkhäuser Boston, 1995.

Chaos & Fractals

Barnsley, Michael F. *Fractals Everywhere.* New York, NY: Academic Press, 1988.
Casti, John L. *Complexification: Explaining a Paradoxical World Through the Science of Surprise.* New York NY: Harper Collins, 1994.
Cohen, Jack and Stewart, Ian. *The Collapse of Chaos: Discovering Simplicity in a Complex World.* New York, NY: Viking Penguin, 1995.
Ekeland, Ivar. *Mathematics and the Unexpected.* Chicago, IL: University of Chicago Press, 1988.
Gleick, James. *Chaos: Making a New Science.* New York, NY: Viking Press, 1987.
Hall, Nina. *Exploring Chaos: A Guide to the New Science of Disorder.* New York NY: W. W. Norton, 1994.
Mandelbrot, Benôit. *The Fractal Geometry of Nature.* San Francisco, CA: W. H. Freeman, 1982.
Peterson, Ivars. *Newton's Clock: Chaos in the Solar System.* New York, NY: W. H. Freeman, 1993.
Peak, David and Frame, Michael. *Chaos Under Control: The Art and Science of Complexity.* New York, NY: W. H. Freeman, 1994.
Peitgen, Heinz-Otto and Richter, Peter H. *The Beauty of Fractals.* New York, NY: Springer-Verlag, 1986.
Peitgen, Heinz-Otto and Saupe, Dietmar. *The Science of Fractal Images.* New York, NY: Springer-Verlag, 1988.

Pickover, Clifford. *Fractal Horizons: The Future Uses of Fractals.* New York, NY: St. Martin's Press, 1996.
Ruelle, David. *Chance and Chaos.* Princeton, NJ: Princeton University Press, 1993.
Schroeder, Manfred. *Fractals, Chaos, Power Laws: Minutes from an Infinite Paradise.* New York, NY: W. H. Freeman, 1995.
Stewart, Ian. *Does God Play Dice? The Mathematics of Chaos.* Oxford: Blackwell, 1989.

History & Culture

Albers, Donald J. and Alexanderson, G.L. *Mathematical People: Profiles and Interviews.* Cambridge, MA: Birkhauser Boston, 1985.
Albers Donald J., et. al. *More Mathematical People: Contemporary Conversations.* Orlando, FL: Harcourt Brace Jovanovich, 1990.
Ascher, Marcia. *Ethnomathematics: A Multicultural View of Mathematical Ideas.* Pacific Grove, CA: Brooks Cole, 1991.
Casti. John L. *Five Golden Rules: Great Theories of 20th-Century Mathematics and Why They Matter.* Wiley, 1995.
Crump, Thomas. *The Anthropology of Numbers.* Cambridge: Cambridge University Pr., 1990.
Dunham, William. *Journey Through Genius: The Great Theorems of Mathematics.* New York, NY: John Wiley & Sons. 1990.
Hardy, G. H. *A Mathematician's Apology.* Cambridge: Cambridge University Pr., 1967.
Hofstadter, Douglas R. *Gödel, Escher, Bach: An Eternal Golden Braid.* New York, NY: Vintage Press, 1980.
Kline, Morris. *Mathematics in Western Culture.* New York, NY: Oxford University Press, 1953.
Kramer, Edna E, *The Nature and Growth of Modern Mathematics.* Princeton, NJ: Princeton University Pr., 1982.
Joseph, George Gheverghese. *The Crest of the Peacock: Non-European Roots of Mathematics.* London: Penguin, 1992.
McLeish, John. *The Story of Numbers: How Mathematics Has Shaped Civilization.* New York NY: Fawcett Columbine, 1994.
Motz, Lloyd. *The Story of Mathematics.* New York, NY: Avon Books, 1995.
Nelson, David, et.al. *Multicultural Mathematics.* New York, NY: Oxford University Press, 1993.
Newman James R. *The World of Mathematics.* (4 Vols.). Microsoft Press, 1988.
White, Alvin. *Essays in Humanistic Mathematics.* Washington, DC: Mathematical Association of America, 1993.

The Mathematical Mind

Changeux, Jean-Pierre and Connes, Alain. *Conversations on Mind, Matter, and Mathematics.* Princeton University Press, 1995.

Davis, Philip J. and Hersh, Reuben. *The Mathematical Experience.* Boston, MA: Birkhauser, 1980.

Guillen, Michael. *Bridges to Infinity: The Human Side of Mathematics.* Boston, MA: Houghton Mifflin, 1983.

Hoffman, Paul. *Archimedes' Revenge: The Joys and Perils of Mathematics.* New York, NY: W.W. Norton & Co., 1988.

Lang, Serge. *The Beauty of Doing Mathematics: Three Public Dialogues.* New York, NY: Springer-Verlag, 1985.

Rothstein, Edward. *Emblems of Mind: The Inner Life of Music and Mathematics.* New York, NY: Random House, 1995.

Nature of Mathematics

Courant, Richard and Robbins, Herbert. *What is Mathematics?* New York: Oxford University Pr., 1960.

Davis, Donald M. *The Nature and Power of Mathematics.* Princeton NJ: Princeton University Press, June 1993.

Davis, Philip J. and Hersh, Reuben. *Descartes' Dream: The World According to Mathematics.* San Diego, CA: Harcourt Brace Jovanovich, 1986.

Devlin, Keith. *Mathematics: The Science of Patterns.* New York, NY: Scientific American Library, 1994.

King, Jerry P. *The Art of Mathematics.* New York: Plenum Press, 1992.

Kitcher, Philip. *The Nature of Mathematical Knowledge.* New York, NY: Oxford University Press, 1983.

Kline, Morris. *Mathematics: The Loss of Certainty.* New York, NY: Oxford University Press, 1980.

Stein, Sherman K. *Mathematics: the Man-Made Universe,* Third Edition. New York, NY: McGraw Hill, 1976.

Stewart, Ian. *Nature's Numbers: The Unreal Reality of Mathematical Imagination.* New York. NY: Basic Books, 1995.

Elements of Mathematics

Barrow, John D. *Pi in the Sky: Counting, Thinking, & Being.* New York. NY: Little, Brown & Co., 1993.

Berlinski, David. *A Tour of Calculus.* New York, NY: Pantheon, 1996.

Devlin, Keith. *Mathematics: The New Golden Age.* London, England: Penguin, 1988.

Eigen, Manfred and Winkler, Ruthild. *Laws of the Game: How the Principles of Nature Govern Chance.* Princeton, NJ: Princeton University Press, 1993.

Ekeland, Ivar. *The Broken Dice and Other Mathematical Tales of Chance.* University of Chicago Press, 1993.

Guillen, Michael. *Five Equations That Changed the World: The Power and Poetry of Mathematics.* New York NY: Hyperion, 1995.

Jacobs Konrad. *Invitation to Mathematics.* Princeton, NJ: Princeton University Press, 1992.

Kline, Morris. *Mathematics and the Search for Knowledge.* New York, NY: Oxford University Press, 1985.

Stewart, Ian. *Concepts of Modern Mathematics.* Mineola, NY: Dover, 1995,

Snapshots & Vignettes

Devlin, Keith. *All the Math That's Fit to Print : Articles from The Manchester Guardian.* Washington, DC: The Mathematical Association of America, 1994.

Dunham, William. *The Mathematical Universe: An Alphabetical Journey Through the Great Proofs, Problems, and Personalities.* New York, NY: John Wiley & Sons. 1994.

Ogilvy. C. Stanley *Excursions in Mathematics.* Mineola, NY: Dover, 1994.

Paulos, John Allen. *Beyond Numeracy: Ruminations of a Numbers Man.* New York, NY: Alfred A. Knopf, 1991.

Peterson, Ivars. *Islands of Truth : A Mathematical Mystery Cruise.* New York, NY: W. H. Freeman, 1990.

Schwartzman, Steven. *The Words of Mathematics.* Washington, DC: Mathematical Association of America, 1994.

Steinhaus, Hugo. *Mathematical Snapshots.* Third Edition. New York, NY: Oxford University Press, 1983.

Stewart, Ian. *The Problems of Mathematics.* Second Edition. New York, NY: Oxford University Press, 1992.

Frontiers of Mathematics

Aczel, Amir D. *Fermat's Last Theorem: Unlocking the Secret of an Ancient Mathematics Problem.* Four Walls Eight Windows, 1996.

Cipra, Barry. *What's Happening in the Mathematical Sciences? Vols. 1-3.* Providence, RI: American Mathematical Society, 1994-96.

Penrose, Roger. *The Emperor's New Mind.* New York, NY: Oxford University Pr., 1989.

Peterson, Ivars. *The Mathematical Tourist: Snapshots of Modern Mathematics.* New York, NY: W. H. Freeman, 1988.

Osserman, Robert, *Poetry of the Universe: The Mathematical Imagination and the Nature of the Cosmos.* New York, NY: Doubleday, 1996.

Steen, Lynn Arthur. *Mathematics Today: Twelve Informal Essays.* New York, NY: Springer-Verlag, 1978.

Innumeracy & Math Anxiety

Buxton, Laurie. *Math Panic.* Portsmouth, NH: Heinemann, 1991.

Dewdney, A. K. *200% of Nothing: An Eye-Opening Tour through the Twists and Turns of Math Abuse and Innumeracy.* New York: John Wiley & Sons. 1993.

Kogelman, Stanley and Heller, Barbara R. *The Only Math Book You'll Ever Need.* Revised Edition. New York NY: Harper Collins, 1995.

Motz, Lloyd, and Weaver, Jefferson Hane. *Conquering Mathematics from Arithmetic to Calculus.* New York, NY: Plenum Press, 1991.

Paulos, John Allen. *Innumeracy : Mathematical Illiteracy and its Consequences.* New York, NY: Vintage Books, 1988.

Ruedy, Elisabeth and Nirenberg, Sue. *Where Do I Put the Decimal Point? How to Conquer Math Anxiety & Let Numbers Work for You.* New York NY: Avon Books, 1992.

Sons Linda R. and Nicholls, Peter J. *Mathematical Thinking in a Quantitative World.* Dubuque, IA: Kendall/Hunt, 1992.

Tobias, Sheila. *Overcoming Math Anxiety.* (Revised Edition). New York, NY: W. W. Norton, 1993.

Tobias, Sheila. *Succeed with Math: Every Student's Guide to Conquering Math Anxiety.* New York, NY: The College Board, 1987.

Public Policy

Crossen, Cynthia. *Tainted Truth: The Manipulation of Fact in America.* New York NY: Simon & Schuster, 1994.

Friedman, Sharon M., and Rogers, Carol L. *Environmental Risk Reporting: The Science and the Coverage.* Bethlehem, PA: Lehigh University, Dept. of Journalism, 1991.

MacNeal, Edward. *Mathsemantics: Making Numbers Talk Sense.* New York, NY: Viking Penguin, 1994.

Paulos, John Allen. *A Mathematician Reads the Newspaper.* New York, NY: Doubleday, 1996.

Porter, Theodore M. *Trust in Numbers: The Pursuit of Objectivity in Science and Public Life.* Princeton, NJ: Princeton University Pr., 1995.

Schwartz, Richard H. *Mathematics and Global Survival,* Second Edition. Needham Heights, MA: Ginn Press, 1991.

Taylor, Alan D. *Mathematics and Politics: Strategy, Voting, Power, and Proof.* New York, NY: Springer-Verlag, 1995.

Education

Mathematical Sciences Education Board. *Reshaping School Mathematics: A Philosophy and Framework for Curriculum.* Washington, DC: National Academy Press, 1990.

Meiring, Steven P., et al. *A Core Curriculum: Making Mathematics Count for Everyone.* Reston, VA: National Council of Teachers of Mathematics, 1992.
NCTM. *Curriculum and Evaluation Standards for School Mathematics.* Reston, VA: National Council of Teachers of Mathematics, 1989.
National Research Council. *Everybody Counts: A Report to the Nation on Mathematics Education.* Washington, DC: National Academy Press, 1989.
Nunes, Terezinha, et. al. *Street Mathematics and School Mathematics.* New York, NY: Cambridge University Press, 1993.
Ohanian, Susan. *Garbage, Pizza, Patchwork Quilts, and Math Magic: Stories about Teachers Who Love to Teach and Children Who Love to Learn.* New York, NY: W. H. Freeman, 1992.
Silver, Edward A., et. al. *Thinking Through Mathematics: Fostering Inquiry and Communication in Mathematics Classrooms.* New York, NY: The College Board, 1995.
Steen, Lynn Arthur. *On the Shoulder of Giants: New Approaches to Numeracy.* Washington, DC: National Academy Press, 1990.
Thiessen, Diane, and Matthias, Margaret. *The Wonderful World of Mathematics: A Critically Annotated List of Children's Books in Mathematics.* Reston, VA: National Council of Teachers of Mathematics, 1992.
Tobias, Sheila. *They're Not Dumb, They're Different: Stalking the Second Tier.* Tucson, AZ: Research Corp., 1990.

Science Literacy

American Association for the Advancement of Science. *Benchmarks for Science Literacy.* New York, NY: Oxford University Press, 1993.
National Research Council. *National Science Education Standards.* National Academy Press, 1996.
Project 2061. *Science for All Americans.* American Association for the Advancement of Science, 1989.
Trefil, James and Hazen, Robert M. *The Sciences: An Integrated Approach.* New York, NY: John Wiley and Sons, 1995.